# Python Programming
# for Mathematics

*Python Programming for Mathematics* focuses on the practical use of the Python language in a range of different areas of mathematics. Through 55 exercises of increasing difficulty, the book provides an expansive overview of the power of using programming to solve complex mathematical problems.

This book is intended for undergraduate and graduate students who already have learned the basics of Python programming and would like to learn how to apply that programming skills in mathematics.

**Features**
- Innovative style that teaches programming skills via mathematical exercises.
- Ideal as a main textbook for Python for Mathematics courses, or as a supplementary resource for Numerical Analysis and Scientific Computing courses.

**Julien Guillod** is an Associate Professor of Applied Mathematics at Laboratoire Jacques-Louis Lions, Sorbonne University, Paris; a part-time member of the Department of Mathematics and Applications, ENS, Paris; and a member of an Inria team. He earned his PhD in Physics from the University of Geneva in 2015.

Guillod's research focuses mainly on the analysis of partial differential equations in fluid mechanics, involving both traditional analysis and numerical simulations. The numerical aspects are mainly used to gain insight into the problems considered, or to discover fundamental properties of the equations studied. His favorite and most commonly used language for these simulations is Python. Most of his research is related in one way or another to the Navier-Stokes equations.

# Chapman & Hall/CRC
# The Python Series

### About the Series

Python has been ranked as the most popular programming language, and it is widely used in education and industry. This book series will offer a wide range of books on Python for students and professionals. Titles in the series will help users learn the language at an introductory and advanced level, and explore its many applications in data science, AI, and machine learning. Series titles can also be supplemented with Jupyter notebooks.

**Image Processing and Acquisition using Python, Second Edition**
*Ravishankar Chityala and Sridevi Pudipeddi*

**Python Packages**
*Tomas Beuzen and Tiffany-Anne Timbers*

**Statistics and Data Visualisation with Python**
*Jesús Rogel-Salazar*

**Introduction to Python for Humanists**
*William J.B. Mattingly*

**Python for Scientific Computation and Artificial Intelligence**
*Stephen Lynch*

**Learning Professional Python Volume 1: The Basics**
*Usharani Bhimavarapu and Jude D. Hemanth*

**Learning Professional Python Volume 2: Advanced**
*Usharani Bhimavarapu and Jude D. Hemanth*

**Learning Advanced Python from Open Source Projects**
*Rongpeng Li*

**Foundations of Data Science with Python**
*John Mark Shea*

**Data Mining with Python: Theory, Applications, and Case Studies**
*Di Wu*

***A Simple Introduction to Python***
*Stephen Lynch*

**Introduction to Python: with Applications in Optimization, Image and Video Processing, and Machine Learning**
*David Baez-Lopez and David Alfredo Báez Villegas*

**Tidy Finance with Python**
*Christoph Frey, Christoph Scheuch, Stefan Voigt and Patrick Weiss*

**Introduction to Quantitative Social Science with Python**
*Weiqi Zhang and Dmitry Zinoviev*

**Python Programming for Mathematics**
*Julien Guillod*

For more information about this series please visit: https://www.routledge.com/Chapman--HallCRC-The-Python-Series/book-series/PYTH

# Python Programming
# for Mathematics

Julien Guillod

## CRC Press

Taylor & Francis Group

Boca Raton  London  New York

CRC Press is an imprint of the
Taylor & Francis Group, an **informa** business

A CHAPMAN & HALL BOOK

Designed cover image: Julien Guillod

First edition published 2025
by CRC Press
2385 NW Executive Center Drive, Suite 320, Boca Raton FL 33431

and by CRC Press
4 Park Square, Milton Park, Abingdon, Oxon, OX14 4RN

*CRC Press is an imprint of Taylor & Francis Group, LLC*

© 2025 Julien Guillod

ISBN: 978-1-032-93338-2 (hbk)
ISBN: 978-1-032-91011-6 (pbk)
ISBN: 978-1-003-56545-1 (ebk)

DOI: 10.1201/ 9781003565451

Typeset in Stix font
by KnowledgeWorks Global Ltd.

*Publisher's note*: This book has been prepared from camera-ready copy provided by the authors.

# Contents

## CHAPTER 11 ■ Differential Equations     166

## CHAPTER 12 ■ Data Science     189

## CHAPTER 13 ■ Cryptography     220

# CHAPTER 1

# Introduction

Python is a leading programming language in the scientific world. It is perfectly adapted to program mathematical problems. This book focuses on the practical use of the Python language in different areas of mathematics: sequences, linear algebra, integration, graph theory, finding zeros of functions, probability, statistics, differential equations, symbolic calculus, and number theory. Through 55 exercises of increasing difficulty, and corrected in detail, it gives a good overview of the possibilities of using programming in mathematics and to be able to solve complex mathematical problems.

It is not necessary to do the exercises in the order suggested, even if some exercises sometimes call upon notions seen in previous exercises. The more difficult exercises are indicated by exclamation marks:

- ! : longer or more difficult;

- !! : quite long and complex;

- !!! : challenge proposed without correction.

This book is the English translation of the second French edition "Programmation Python par la pratique" published by Dunod in 2024. These exercises are used as a basis for the practical work given at Sorbonne University in the framework of the undergraduate mathematics studies.

The complete source code of the book is available online at the address: `http s://python.guillod.org/`. This site is updated regularly, so it may differ from this book in the future.

## Acknowledgments

Thanks to Marie Postel and Nicolas Lantos for their careful proofreading of the first version of this manuscript and for many corrections and suggestions. Special thanks to Johann Faouzi and Louis Thiry.

Thanks also to the members of the Sorbonne University pedagogical teams who used these exercises for their feedback and contributions: Mathieu Barré,

Constantin Bône, Jules Bonnard, Cédric Boutillier, Thibault Cimic, Jeanne Decayeux, Cécile Della Valle, Guillaume Duboc, Jean-Jil Duchamps, Jean-Merwan Godon, Elise Grosjean, Cindy Guichard, Sidi-Mahmoud Kaber, Nicolas Lantos, Mathieu Mari, David Michel, Leo Miolane, Anouk Nicolopoulos, Arnaud Padrol, Diane Peurichard, Marie Postel, Xavier Poulot-Cazajous, Alexandre Rege, Othmane Safsafi, Emmanuel Schertzer, Agustín Somacal, Didier Smets, Robin Strudel, Gauthier Tallec, Nicolas Thomas, Paul Vernhet, Jules Vidal, and Raphaël Zanella.

Finally, I would like to thank the students who worked on these exercises for their constructive feedback, which contributed to the improvement of this collection.

The attentive reader is thanked in advance for pointing out typos or other errors.

## 1.1  WHY PYTHON?

Python is a general-purpose interpreted programming language that has the particularity of being very readable and pragmatic. It has a very large base of external modules, especially scientific ones, which makes it particularly attractive for programming mathematical problems. The fact that Python is an interpreted language makes it slower than compiled languages, but it ensures a great speed of development which allows humans to work a little less while the computer has to work a little more. This particularity makes Python one of the main programming languages used by scientists.

## 1.2  PREREQUISITES

This book does not aim to explain the syntax and principles of the Python language, so the prerequisite is to know the basics. There are many resources to update yourself if needed, for example:

- the online course *Python Programming MOOC* by the University of Helsinki (https://programming-22.mooc.fi/);

- the book *Python for Everybody* by Charles R. Severance (https://www.py 4e.com/);

- various online courses, like *Crash Course on Python* (https://www.course ra.org/learn/python-crash-course).

Moreover, the realization of the exercises requires access to a computer or an online service with Python 3.6 (or more recent) completed by the following modules: NumPy, SciPy, SymPy, Matplotlib, Numba, NetworkX, and Pandas. The use of a code editor allowing writing in Python is also highly recommended. It is suggested here to use Jupyter Lab, which allows both the writing of interactive notebooks and scripts and also the addition of one's own solutions below the statements, which is very practical. It is not necessary to use Jupyter Lab, other environments are also suitable, such as Spyder or Jupyter Notebook.

The following sections describe how to install and run the Python environment or use it online without installation.

## 1.3 DOCUMENTATION

It is generally not useful (nor desirable) to know all the functions and subtleties of the Python language for occasional use. However, it is essential to know how to use the documentation efficiently. The official documentation is available at the `https://docs.python.org/`. The language and version can be selected in the upper left corner. It is strongly recommended to look at how the documentation is written and to learn how to use it.

## 1.4 INSTALLATION

People who cannot or do not want to install Python can go directly to section 1.5 for alternatives available online without installation.

There are basically four ways to install Python and the modules required to perform the exercises:

- **Anaconda** is a complete Python distribution, *i.e.,* it directly installs a very large quantity of modules (much more than necessary to do the following exercises). The advantage of this installation is that it is very simple; the disadvantage is that it takes a lot of disk space. This is the preferred method if you are running Windows or MacOS and do not have disk space problems.

- **Miniconda** is a lightweight version of Anaconda, which by default installs only the base. The advantage is that it takes up little disk space, but it requires an additional action to install the modules required to do the exercises. This is the preferred method if you are running Windows or MacOS and have little disk space available.

- **Linux repositories:** Most Linux distributions allow you to install Python and the core modules directly from the package repositories that come with them. This is the preferred method under Linux.

- **Pip** is a package manager for Python. This is the preferred method to add a module if Python is already installed by your operating system, and this module is not included in the packages of your distribution. This method allows a more detailed and advanced management of installed modules than what is proposed with the previous methods.

**Installation with Anaconda:** The easiest way to install Python 3 and all necessary dependencies on Windows and MacOS is to install Anaconda. The disadvantage of Anaconda is that its installation takes a lot of disk space because many modules are installed by default. Detailed installation procedures for each operating system are described at: `https://docs.anaconda.com/anaconda/install/`. In summary, the installation procedure is as follows:

1. Download Anaconda for Python 3 from the address: `https://www.anacon da.com/download`.

2. Double-click on the downloaded file to launch the installation of Anaconda, then follow the installation procedure (it is not necessary to install VS Code).

**Installation with Miniconda:** The Miniconda distribution has the advantage over Anaconda of taking up little disk space at the cost of having to install the necessary modules manually. The quick installation procedure is as follows:

1. Download Miniconda for Python 3 from the address: `https://docs.anaco nda.com/miniconda/`.

2. Double-click on the downloaded file to launch the installation of Miniconda, then follow the installation procedure.

3. Once the installation is complete, launch Anaconda Prompt from the Start menu or from the list of applications.

4. In the terminal, type the command:

```
─────── Terminal ───────
conda install numpy scipy sympy matplotlib numba
 ↳  networkx pandas jupyterlab
```

and confirm the installation of the dependencies.

5. Optionally (but recommended), install the LSP (Language Server Protocol) interface with the commands:

```
─────── Terminal ───────
conda config --append channels conda-forge
conda install jupyterlab-lsp python-lsp-server
```

**Installing from repositories:** Most Linux distributions allow to easily install Python and the most standard modules directly from the distribution repositories. The following procedure is for Ubuntu, but can be easily adapted to other distributions.

1. Install Python 3:

```
─────── Terminal ───────
sudo apt install python3 python3-pip
```

2. Update Pip:

```
                          ─ Terminal ─
pip install --upgrade pip
```

3. Install the modules NumPy, SciPy, SymPy, Matplotlib, Numba, NetworkX, and Pandas:

```
                          ─ Terminal ─
sudo apt install python3-numpy python3-scipy
  ↵  python3-sympy python3-matplotlib python3-numba
  ↵  python3-networkx python3-pandas
```

4. Jupyter Lab is not available in the Ubuntu packages, so it must be installed with Pip:

```
                          ─ Terminal ─
pip install jupyterlab
```

5. Optionally (but recommended), install the LSP (Language Server Protocol) interface with the command:

```
                          ─ Terminal ─
pip install --upgrade jupyterlab-lsp
  ↵  python-lsp-server[all]
```

See the following remark on Pip if this last command does not work.

**Advanced installation with Pip:** The following procedure describes the manual installation of modules with the Pip manager.

1. Install Python from the address: https://www.python.org/downloads/.

2. Install Pip from the address: https://pip.pypa.io/en/stable/installation/.

3. Install the required modules by typing the following command line in a terminal:

```
                          ─ Terminal ─
pip install numpy scipy sympy matplotlib numba networkx
  ↵  pandas jupyterlab
```

4. Optionally (but recommended), install the LSP (Language Server Protocol) interface with the command:

```
                          ─ Terminal ─
pip install jupyterlab-lsp python-lsp-server[all]
```

**Remark:** Depending on the operating system, the command `pip` must be replaced by `pip3`. If you encounter a permissions problem when executing these commands, you should probably add `--user` to the end of the previous command.

## 1.5 LAUNCH OF JUPYTER LAB

**With Anaconda Navigator:** If Anaconda Navigator has been installed (as is the case with Anaconda), simply launch Anaconda Navigator from the start menu or application list, then click on the "jupyterlab" icon.

**On the command line:** With Anaconda or Miniconda, launch Anaconda Prompt from the Start menu or application list. In other cases, simply open a terminal (if a virtual environment has been created, don't forget to activate it). To launch Jupyter Lab from the command line, type `jupyter lab` in the terminal. To quit, click on `Shutdown` in the `File` menu of the Jupyter Lab window. It is also possible to type `Ctrl+C` followed by `y` in the terminal where the command `jupyter lab` was executed.

**Online without installation:** For people who cannot or do not want to install Python and the necessary dependencies on their own computer, it is possible to use Jupyter Lab online with GESIS: `https://notebooks.gesis.org/binder/v2/gh/guillod/python-book/HEAD`. No account is required, but modified documents are automatically deleted on exit, so it's essential to save them on your own computer before leaving. Otherwise, various services offer the possibility to use Jupyter Lab for free after creating an account:

- CoCalc (`https://cocalc.com/`)

- Google Colaboratory (`https://colab.research.google.com/`)

## 1.6 USE OF JUPYTER LAB

Once Jupyter Lab is launched, the window shown in Figure 1.1 should appear in a browser.

Jupyter Lab essentially allows us to process three types of documents: **notebooks**, **scripts**, and **terminals**. A notebook consists of cells that can contain either code or text in Markdown format. Code cells can be evaluated interactively on demand, which allows great flexibility. Text cells can contain comments, titles, or LaTeX formulas as represented in Figure 1.2.

A Python script is simply a text file containing Python instructions. It is executed in its entirety from A to Z and it is not possible to interact interactively with it during its execution (unless it has been explicitly programmed). To execute a Python script, it is necessary to open a terminal.

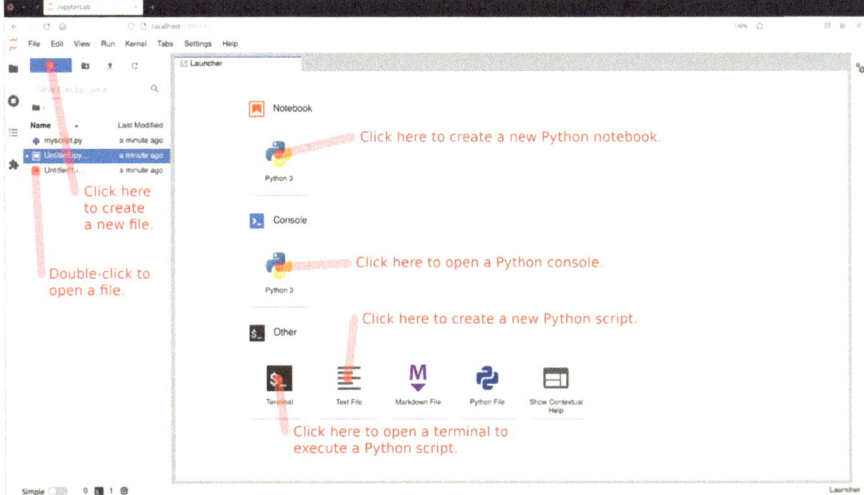

Figure 1.1   Jupyter Lab launch window.

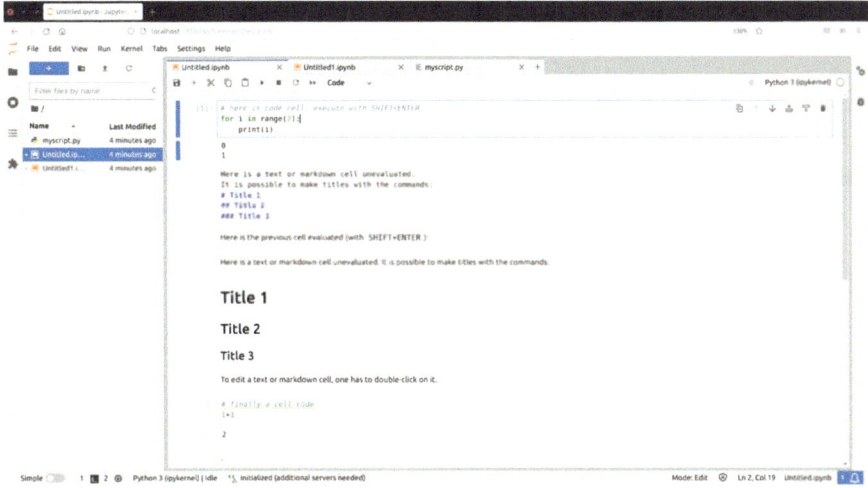

Figure 1.2   A Jupyter notebook is composed of several cells. Here, a code cell is followed by an unevaluated text cell, then the same cell, evaluated, and finally a last code cell.

**Basic commands:**

- Create a new file: click on the "+" button on the top left, then choose the type of file to create.

- Rename a file: click with the second mouse button on the title of the notebook (either in the tab or in the file list).

- Change cell type: drop-down menu to choose between "Code" and "Markdown".

- Execute a code cell: combination of keys SHIFT+ENTER.

- Format a text cell: combination of keys SHIFT+ENTER.

- Edit a text cell: double-click on the cell.

- Run a script: type python scriptname.py in a terminal to run the script scriptname.py.

- Rearrange cells: click and drop.

- Juxtaposing tabs: click and drop.

The detailed documentation of Jupyter Lab is available at the address: https://jupyterlab.readthedocs.io/.

## 1.7 ADVANCED USE OF JUPYTER LAB

Recent versions of Jupyter Lab (3 and higher with ipykernel greater than 6) feature a particularly useful debugger and LSP (Language Server Protocol) interface. The debugger allows you to find errors in the code by stopping the program at particular points to understand what's going on. Documentation and a tutorial on how to use the debugger are available at the address: https://jupyterlab.readthedocs.io/en/stable/user/debugger.html. The LSP interface provides access to documentation and function signatures, offers code diagnostics and autocompletion. Information on installing and using the LSP interface is available at the address: https://github.com/jupyter-lsp/jupyterlab-lsp/blob/master/README.md.

**Debugger:** When writing code, it's natural to make mistakes, and one important aspect is to locate and identify them efficiently. To do this, it's possible to put print commands in the right places, but it's more appropriate to use a debugger for this. To activate Jupyter Lab's debugger, click on the beetle in the top right-hand corner so that it turns orange. When the debugger is activated, the list of global variables is available in the dedicated bar. The most useful aspect of the debugger is the definition of breakpoints, which allow you to execute the code up to a certain line and inspect the state of the program at that point. To do this, consider the following function, which adds two numbers:

```python
def add(a, b):
    res = a + b
    return res
```

Clicking to the left of a code line number places a breakpoint indicated by a red dot. Here, we propose to click on the second line performing the addition. By executing the following function call code:

```
resultat = add(1, 2)
print(resultat)
```

the program will stop at the second line of the add function. You can view the values of variables a and b in the "Variables" tab and the relevant source code in the "Sources" tab. Breakpoints are grouped together in the "Breakpoints" tab. By navigating the "Callstack" tab, you can continue program execution up to the next breakpoint.

**Hover:** When hovering over any part of the code with the mouse, if a part of the code becomes underlined it is then possible to get information about the function with the CTRL key. For example, by hovering the mouse over the following code:

```
from numpy import linalg
```

and by pressing the CTRL key with the mouse on numpy or linalg a window with explanations about these modules is displayed. This is also the case for manually defined functions if they contain a docstring:

```
def square(x):
    """Definition of the function x -> x^2"""
    return x*x
```

Moving the mouse over the word square:

```
r = square(4)
```

underlines it, and with the key CTRL the definition appears.

**Diagnostics:** Critical errors or warnings are indicated by an underline in red or orange, for example, in the case of an undefined variable:

```
def f(x):
    if x:
        undefined_variable
    return x
```

**Suggestions:** By typing linalg. in a cell, suggestions of functions available in this module are displayed. In other cases, the suggestions are activated with the TAB key. This is the case, for example, with a manually defined dictionary:

```
dic = {'key1':3, 'key2':5}
```

By typing dic[ in a cell followed by the TAB key, the suggestions 'key1' and 'key2' come up.

**Signatures:** By typing linalg.solve( then comes the help and signature of this function, *i.e.*, the way the arguments are to be used in this function. By placing the mouse on the word solve with the CTRL key, there comes also a description of the function.

**References:** By clicking on a symbol, its other uses are highlighted.

**Definition:** By clicking with the right mouse button on a symbol and then on "Jump to definition", it is possible to go to the definition of the function in question. It is possible to test on the following code, for example:

```
f(None)
```

**Renaming:** It is possible to rename a variable intelligently (*i.e.*, without renaming local variables, for example) by right-clicking on the variable in question and selecting "Rename symbol".

**Diagnostics panel:** It is possible to sort and navigate through the diagnostics using the "Diagnostics panel". To open it, simply select "Show diagnostics panel" from the context menu of a cell (right mouse button).

**Personalization:** The "Settings" menu of Jupyter Lab allows you to customize the working environment, especially to choose the theme, the font size, the default indentation, but also many other more advanced options.

# CHAPTER 2

# Data Structures

To represent data structures, Python offers four basic types: lists (type `list`), tuples (type `tuple`), sets (type `set`), and dictionaries (type `dict`). The purpose of this chapter is to show the fundamental differences between these data structures and to explain what they are best suited for. Detailed documentation on data structures is available at the address: `https://docs.python.org/3/tutorial/datastructures.html`.

## Concepts covered

- data structures (list, tuple, set, dictionary)

- mutable and immutable types

- hashable type

- list, set and dictionary comprehensions

- numerical series

DOI: 10.1201/9781003565451-2

# EXERCISES

## EXERCISE 2.1   LISTS

A list is a structure allowing to store heterogeneous elements:

```python
list0 = [0, 5.4, "string", True]
```

Lists are mutable, *i.e.*, it is possible to modify an element, add one or delete one, without having to redefine the whole list.

```python
list0[3] = False # replace True by False
list0.append("new") # add the string "new" to the list
list0.insert(2, 34) # insert 34 in place of 2
list0.remove(0) # remove 0
```

In particular, care must be taken when copying a list. If we execute the following code:

```python
list1 = list0
list1[2] = "change"
list0
```

then `list0` is also modified and is equal to `list1`. To create a real copy, you have to use the following code:

```python
list2 = list0.copy()
list2[2] = "rechange"
list0
```

which does not modify `list0`. Note that it is possible to modify the elements of a list inside a function:

```python
def f(l):
    l[0] = 0
f(list0)
```

Finally, it is possible to create lists with the help of list comprehension:

```python
list1 = [2*i+1 for i in range(10)]
```

**a.** Search the documentation for the syntax to concatenate two lists.
*Hint: See the documentation at the address:* `https://docs.python.org/3/libr ary/stdtypes.html#sequence-types-list-tuple-range`.

**b.** Look in the documentation for the syntax to extract a slice from a list, *i.e.*, if a is, for example, a list of length 10, return the elements from 6 to 9.

*Hint: See the documentation at the address:* `https://docs.python.org/3/libr`
`ary/stdtypes.html#sequence-types-list-tuple-range`.

**c.** Search the documentation for the syntax to return the length of a list.

**d.** Write a function `fibonacci(N)` that returns the list of $N$ first terms of the Fibonacci sequence defined by $u_{n+2} = u_{n+1} + u_n$ with $u_0 = 0$ and $u_1 = 1$.

**e.** Write a function `pascal(N)` that returns the $N$-th line of Pascal's triangle:

$$\begin{array}{ccccccccccc}
&&&&& 1 \\
&&&& 1 && 1 \\
&&& 1 && 2 && 1 \\
&& 1 && 3 && 3 && 1 \\
& 1 && 4 && 6 && 4 && 1 \\
1 && 5 && 10 && 10 && 5 && 1
\end{array}$$

**f.** Let $(u_n)_{n\in\mathbb{N}}$ and $(v_n)_{n\in\mathbb{N}}$ be the sequences defined by $u_0 = 1$, $v_0 = 1$, and

$$u_{n+1} = u_n + v_n, \qquad\qquad v_{n+1} = 2u_n - v_n,$$

for $n \geq 0$. Calculate $u_{100}$ and $v_{100}$.
*Answer:* $u_{100} = v_{100} = 717897987691852588770249$.

**g.** Write a function `vk(n0,K)`, which for two integers $n_0$ and $K \geq 1$ computes the sequence of values $v_k$ defined by $v_0 = n_0$ and

$$v_{k+1} = \begin{cases} 3v_k + 1 & \text{if } v_k \text{ is odd,} \\ \dfrac{v_k}{2} & \text{if } v_k \text{ is even,} \end{cases}$$

for $0 \leq k < K$. For $K = 1\,000$ and various values of $n_0 \in \{10, 100, 1\,000, 10\,000\}$, display the last five calculated values, *i.e.*, $(v_{K-4}, v_{K-3}, v_{K-2}, v_{K-1}, v_K)$.
*Answer: The following statements are true:*

```
vk(10,1000)    == [1, 4, 2, 1, 4]
vk(100,1000)   == [2, 1, 4, 2, 1]
vk(1000,1000)  == [1, 4, 2, 1, 4]
vk(10000,1000) == [4, 2, 1, 4, 2]
```

## EXERCISE 2.2   TUPLES

Tuples allow, just like lists, to store heterogeneous elements:

```
tuple0 = (0, 5.4, "string", True)
```

But unlike lists, tuples are not mutable. It is not possible to modify an element, add one or delete one, without redefining the whole tuple. The advantage of a tuple over a list is that it is hashable, *i.e.*, it can be used as a key in a dictionary. Finally, it is possible to assign variables inside a tuple, for example:

```
(a,b) = (1,9)
```

This is especially useful for exchanging two variables without having to use an additional variable:

```
(a,b) = (b,a)
```

**a.** Check that a tuple is immutable.

**b.** Define a function mdlast(lst,val) having as argument a list of integer tuples lst and an integer val and return the list of tuples with the last element of each tuple replaced by val. For example, if lst = [(10, 20), (30, 40, 50, 60), (70, 80, 90)], then mdlast(lst,100) should return [(10, 100), (30, 40, 50, 100), (70, 80, 100)].

**c.** How to convert a tuple into a list and vice versa?

## EXERCISE 2.3   SETS

Sets are used to store heterogeneous elements in the mathematical sense of set theory:

```
set0 = {0, 5.4, "string", True}
```

It is possible to test if an element belongs to a set:

```
if "string" in set0:
    print("inside")
```

Sets are mutable, so it is possible to add or remove an element from a set:

```
set0.add(18) # add 18 to the set
set0.add(0) # add 0 to the set (this does nothing as 0
 ↵    already belongs to the set)
set0.remove("string") # remove "string to the set
```

On the other hand, sets can only contain hashable elements, *i.e.*, immutable. In particular a set cannot contain another set:

```
set1 = {{1,2},{3},{4}}
TypeError: unhashable type: 'set'
```

Note that in Python there are also immutable sets, called frozenset:

```
frozenset0 = frozenset([0, 5.4, "string", True])
```

A string can be transformed into a set:

```python
set1 = set('abracadabra')
```

As with lists, it is possible to make set comprehensions:

```python
set2 = {x for x in 'abracadabra' if x not in 'abc'}
```

In this example, strings are automatically transformed into sets. Note that the empty set is defined by `set()`.

**a.** Define a function `divisible(n)` that returns the set of integers divisible by n less than or equal to 100.

**b.** Search the literature to find the intersection, union, and difference of two sets. Determine the numbers less than or equal to 100 that are not divisible by 2 but divisible by 3 and 5.
*Hint: See the documentation of* `set` *at the address:* `https://docs.python.org/3/library/stdtypes.html#set`.


## EXERCISE 2.4   DICTIONARIES

Dictionaries are a structure allowing to store heterogeneous elements indexed by keys (also heterogeneous):

```python
dict0 = {"apples": 0, "pears": 4, 12: 2}
```

The elements of a dictionary are accessible through the keys:

```python
dict0["apples"]
dict0[12]
```

A dictionary can be seen as an associative array associating to each key a value. The list of keys and the list of values are accessible, respectively, with `dict0.keys()` and `dict0.values()`. Dictionaries are mutable, so it is possible to modify a key-value association and to add or remove one:

```python
dict0["apples"] = 3 # modify the value associated to apples
dict0["oranges"] = "many" # add oranges as key with value
    "many"
del dict0["pears"] # remove the key pears, hence the value
dict0.pop("apples") # remove the key apples, hence the value
```

Although a dictionary is mutable, the keys that compose it must be hashable objects, *i.e.*, immutable. Thus, a list or a set cannot be used as keys in a dictionary:

```python
dict0[list0] = "test"
TypeError: unhashable type: 'list'
```

```
dict0[set0] = "retest"
TypeError: unhashable type: 'set'
```

On the other hand, it is possible to have a tuple or a frozenset as a key:

```
dict0[tuple0] = "test"
dict0[frozenset0] = "rest"
```

hence the interest of frozensets. As for lists and sets, it is possible to make dictionary comprehensions:

```
dict1 = {x: x**2 for x in range(5)}
```

Finally, an interesting thing about dictionaries is the unpacking illustrated by the following example:

```
def add(a=0, b=0):
    return a + b
d = {'a': 2, 'b': 3}
add(**d)
```

**a.** How to define an empty dictionary?

**b.** How to concatenate several dictionaries together?

**c.** We consider a list of words:

```
words = ['Apricot', 'Cranberry', 'Pineapple', 'Banana',
    'Blackcurrant', 'Cherry', 'Lemon', 'Clementine',
    'Quince', 'Date', 'Strawberry', 'Raspberry',
    'Pomegranate', 'Gooseberry', 'Persimmon', 'Kiwi',
    'Litchi', 'Mandarin', 'Mango', 'Melon', 'Mirabelle',
    'Nectarine', 'Orange', 'Grapefruit', 'Papaya', 'Peach',
    'Pear', 'Apple', 'Plum', 'Grape']
```

Write a function position(words, x, n) that returns the list of words with the character x as their n-th letter (starting from zero, as in Python).
*Answer: For example,* position(words,'e',4) *should return the list:*

```
['Clementine', 'Gooseberry', 'Grapefruit', 'Apple', 'Grape']
```

**d.** Assuming that the list of words is very long, then each time the position function is evaluated, the whole set of words is searched, which takes quite a long time. To improve this, build a dictionary mots_dict having as keys the tuples (x,n) and as values the list of words having the character x as n-th letter, *i.e.*, such that mots_dict[x,n] returns the same thing as position(words, x , n) except for the order. Thus, the words list is traversed only once during dictionary construction and then dictionary evaluation is extremely fast for any query.

# SOLUTIONS

## SOLUTION 2.1   LISTS

**a.** The + operator allows the concatenation of lists:

```
list1 = [1,3,4,"a",29]
list2 = ["e",37,2]
list1 + list2
```

**b.** You have to use `list` to transform the type `range` into a list:

```
a = list(range(10))
a[6:10]
```

**c.** The `len` function returns the length of a list:

```
list3 = [2,5,"19",4,8,"R"]
len(list3)
```

**d.** The idea is to build a list element by element from the first two elements:

```
def fibonacci(N):
    # initialization
    out = [0,1]
    for i in range(2,N):
        # add to the list out the sum of the two previous
        ↵   element
        out.append(out[i-1]+out[i-2])
    return out
```

**e.** By using a recursive function, the problem is reduced to writing the logic to move from one line to another:

```
def pascal(N):
    # initialization for the recursion
    if N == 1:
        return [1]
    else:
        # previous line
        previous = pascal(N-1)
        # return the new line composed in terms of the
        ↵   preceding line
        return [1] + [previous[i] + previous[i+1] for i in
        ↵   range(N-2)] + [1]
```

**f.** The following trick allows us to set the new values of u and v at once, otherwise you have to use an additional variable to save one of the old values:

```python
def uv(n):
    # initial values
    u = 1
    v = 1
    for i in range(n):
        # this allows to define the new u and v
        ↵   simultaneously
        u,v = u+v, 2*u-v
    return [u,v]
print(uv(100))
```

**g.** Since only the last five values are requested, it is useless to store them all. As only even numbers are divided by 2, the sequence remains a sequence of integers. To keep this property in Python, you have to use the integer division:

```python
def vk(n0,K):
    # initialization
    v = int(n0)
    # list to store the last 5 results
    out = []
    for i in range(K):
        # if vk even
        if v % 2 == 0:
            v = v//2
        else:
            v = 3*v+1
        # return the last 5 values
        if i in range(K-5,K):
            out.append(v)
    return out
for n0 in [10,100,1000,10000]:
    print(vk(n0,1000))
```

## SOLUTION 2.2   TUPLES

**a.** If you try to modify a tuple, an error appears:

```python
tuple0 = (0, 5.4, "string", True)
tuple0[3] = False # impossible to modify a value
TypeError: 'tuple' object does not support item assignment
del tuple0[1] # impossible to delete a value
TypeError: 'tuple' object doesn't support item deletion
```

**b.** Since tuples are immutable, a simple way to do this is to select all but the last element of the tuple and concatenate the tuple to an element (100,):

```
def mdlast(lst,val):
    return [t[:-1] + (100,) for t in lst]
lst = [(10, 20), (30, 40, 50, 60), (70, 80, 90)]
mdlast(lst,100)
```

**c.** The operators `list` and `tuple` allow to make conversions to their respective types:

```
list0 = [0, 5.4, "string", True]
tuple0 = (0, 5.4, "string", True)
list0 == list(tuple0)
tuple0 == tuple(list0)
```

## SOLUTION 2.3   SETS

**a.** The set understanding allows to simply return the integers that are multiples of n:

```
def divisible(n):
    return {n*i for i in range(100//n+1)}
```

**b.** We calculate the intersection of the numbers divisible by 3 and 5, then we subtract the numbers divisible by 2:

```
inter = divisible(3).intersection(divisible(5)) # numbers
  ↳ divisible by 3 and by 5
inter = inter.difference(divisible(2))
inter
```

or equivalent:

```
( divisible(3) & divisible(5) ) - divisible(2)
```

## SOLUTION 2.4   DICTIONARIES

**a.** There are two ways to define an empty dictionary, either with `dict()` or with `{}`:

```
empty_dict = dict()
empty_dict = {}
```

**b.** One method is to create an empty dictionary and then add data from other dictionaries to it:

```
dic1={1:10, 2:20}
dic2={2:20, 3:30, 4:40}
dic3={5:50, 6:60, 3:99}
out = {}
for d in (dic1, dic2, dic3): out.update(d)
```

It is also possible to use unpacking:

```
out = {**dic1, **dic2, **dic3}
```

or from Python 3.9:

```
out = dic1 | dic2 | dic3
```

**c.** The idea is to fill in a list as you go along, going through all the words:

```
def position(words, x, n) :
    out = [ ]
    # iterate over the words
    for word in words :
        # add the word if the nth letter exists and is x
        if n < len(word) and word[n] == x :
            out.append(word)
    return out
print(position(words,'e',4))
```

We note here that in a condition formed by the conjunction of several predicates, as soon as one predicate is false, the following ones are not executed. The order of the predicates is therefore important, for example, the previous code returns an error with if word[n] == x and n < len(word):.

**d.** The use of get allows an optional parameter to be returned when trying to access a key that does not exist:

```
words_dict = {}
for word in words:
    for i,c in enumerate(word) :
        words_dict[(c,i)] = words_dict.get((c,i), []) +
        ↵    [word]
print(words_dict['e',4])
```

# CHAPTER 3

# Homogeneous Structures

Python's default data structures can handle heterogeneous data (for example, integers and strings). This feature makes Python data structures extremely flexible at the expense of performance. Indeed, since heterogeneous data must be supported, it is not possible to allocate a fixed memory range for a data structure, which slows down its use. Particularly in mathematics, homogeneous datasets of fixed size (list of integers, real or complex vectors, matrices...) appear very regularly. The NumPy module defines the `ndarray` type that is optimized for such homogeneous data structures of fixed sizes. The NumPy documentation is available at the address: `https://numpy.org/doc/stable/`.

To load the NumPy module, it is usual to proceed as follows:

```
import numpy as np
```

## Concepts covered

- homogeneous data table

- slicing

- vector operations

- indexing and selection

DOI: 10.1201/9781003565451-3

# EXERCISES

## EXERCISE 3.1   INTRODUCTION TO NUMPY

**Creation:** The size and type of the elements of a NumPy array must be known in advance. The first way to create a NumPy array is to construct an array filled with zeros by specifying the size and type:

```
array0 = np.zeros(3, dtype=int) # vector of 3 integers
array1 = np.zeros((2,4), dtype=float) # array of floats of
 ↳  size 2x4
array2 = np.zeros((2,2), dtype=complex) # square matrix of
 ↳  complex numbers of size 2x2
array3 = np.zeros((5,6,4)) # three-dimensional array of
 ↳  floats
```

The second way is to pass the data directly:

```
array4 = np.array([1,4,5]) # vector of integers
array5 = np.array([[1.1,2.2,3.3,4.4],[1,2,3,4]]) # matrix of
 ↳  floats of size 2x4
array6 = np.array([[1+1j,0.4],[3,1.5]]) # matrix of complex
 ↳  numbers of size 2x2
```

NumPy will then determine itself the type and the size of the array. Note that it is possible to force the type:

```
array0 = np.array([1,4,5], dtype=complex) # vector of complex
 ↳  numbers
```

The type of the elements of the NumPy array `array1` can be determined by `array1.dtype`. The size of this array is given by `array1.shape`. The following commands are used to access the array elements:

```
array4[1] # return 4
array5[1,3] # return 4.0
```

Note that the indices start at 0 and not at 1. NumPy arrays are mutable in the sense that the data can be modified while keeping the same type and size:

```
array0[1] = 4
array1[1,3] = 3.3
array3[3,4,2] = 3
```

**Slicing:** Slicing allows you to access certain parts of a table:

```
array4[2:3] # return the elements of indices between 2 and 3
array1[0,:] # return the first row of array1
array1[:,-1] # return the last column of array1
array3[3,3:5,1:4] # return the corresponding sub-matrix
```

**Iteration:** It is possible to iterate an array on its first dimension, for example, to return the sum of the rows:

```
for i in array5:
    print(np.sum(i))
```

**a.** Study the documentation for the arange function and use this function to generate the vectors $(5, 6, 7, 8, 9)$ and $(3, 5, 7, 9)$.
*Hint: The documentation of the* arange *function is available at the address:* https://numpy.org/doc/stable/reference/generated/numpy.arange.html.

**b.** Study the documentation for the function linspace and use it to generate 10 evenly spaced numbers in the interval $[2, 5]$.

**c.** Read the documentation for the reshape function and perform the following transformations in succession:

$$(1, 2, 3, 4, 5, 6) \rightarrow \begin{pmatrix} 1 & 2 \\ 3 & 4 \\ 5 & 6 \end{pmatrix} \rightarrow \begin{pmatrix} 1 & 2 & 3 \\ 4 & 5 & 6 \end{pmatrix} \rightarrow \begin{pmatrix} 1 & 4 \\ 2 & 5 \\ 3 & 6 \end{pmatrix}$$

## EXERCISE 3.2   OPERATIONS ON ARRAYS

The basic arithmetic operations on NumPy arrays are performed element by element:

```
mat1 = np.array([[1,2.5,3],[5,6.1,8],[3,2,5]])
mat2 = np.array([[1,0.5,0],[0,0.9,8],[2,0,0]])
mat1 + mat2 # return the sum element by element
mat1 * mat2 # return the product element by element (not the
    ↳  matrix product)
10*mat1**2 # return 10 times the square of the elements of
    ↳  mat1
```

Most of the mathematical functions defined by NumPy (see https://numpy.org/doc/stable/reference/routines.math.html) are also performed element by element:

```
np.cos(mat1) # return the cosine element by element of mat1
np.exp(mat1) # return the exponential element by element of
    ↳  mat1
```

The matrix product can be performed in one of three ways:

```
np.dot(mat1,mat2)
mat1.dot(mat2)
mat1 @ mat2
```

**a.** Given a vector $(v_0, v_1, \dots, v_{n-1})$, the discrete derivative of this vector is defined by the vector $(d_0, d_1, \dots, d_{n-2})$ given by $d_i = v_{i+1} - v_i$ for $i = 0, 1, \dots, n - 2$. Write a function `diff_list` that computes the discrete derivative of a list and a function `diff_np` that does the same operation but on NumPy vectors using slicing.

**b.** Let a_list and a_np be, respectively, a list and an array of 1 000 elements drawn at random in the interval $[0, 1]$:

```
a_list = [np.random.random() for _ in range(1000)]
a_np = np.random.random(1000)
```

Compare the execution time of `diff_list(a_list)` and `diff_np(a_np)`.
*Hint: In Jupyter Lab, it is very easy to determine the time taken by a cell to evaluate itself, just start the cell with %%time, for example:*

```
%%time
result = diff_list(a_list)
```

*To evaluate the cell multiple times and average the execution time to get a more accurate result, replace %%time with %%timeit. The documentation is available at the address:* `https://ipython.readthedocs.io/en/stable/interactive/magics.html#magic-timeit`.
*Answer: The execution time with NumPy tables should be approximately 50 to 100 times faster than with lists!*

## EXERCISE 3.3   VANDERMONDE MATRIX

For $p, n \in \mathbb{N}^*$ and $\mathbf{x} = (x_1, \dots, x_p)$ a vector of size $p$, the corresponding Vandermonde matrix is defined by:

$$
V(\mathbf{x}, n) = \begin{pmatrix}
1 & x_1 & x_1^2 & \cdots & x_1^{n-1} & x_1^n \\
1 & x_2 & x_2^2 & \cdots & x_2^{n-1} & x_2^n \\
\vdots & \vdots & \vdots & \ddots & \vdots & \vdots \\
1 & x_{p-1} & x_{p-1}^2 & \cdots & x_{p-1}^{n-1} & x_{p-1}^n \\
1 & x_p & x_p^2 & \cdots & x_p^{n-1} & x_p^n
\end{pmatrix}.
$$

**a.** Write a function that constructs the matrix $V(\mathbf{x}, n)$ element by element using a double loop.

**b.** After establishing a relationship to write the $k$-th column of $V(x, n)$ solely as a function of $x$ and $k$, write a second function that constructs the matrix $V(x, n)$ column by column using this relationship.

**c.** After establishing a relationship between the $k$-th column of $V(x, n)$, its $(k-1)$-th column, and the vector $x$, write a third function that constructs the matrix $V(x, n)$ column by column using this relationship.

**d.** Compare the execution times of these three functions for $n = 150$, $p = 100$, and $x$ generated randomly.

## EXERCISE 3.4 ARRAY INDEXING (!)

Slicing allows you to select blocks in an array, but it is also possible to select disparate elements using an array as indexing:

```
a = np.arange(12)**2 # array of perfect squares
i = np.array([1,3,8,5]) # array of indices
a[i] # array of the elements with indices i
```

Note that it is also possible to index by an array of higher dimension. The result is then an array of the same shape as the index:

```
j = np.array([[3,4],[9,7]]) # two-dimensional array of
 ↵   indices
a[j] # select the elements with indices j
```

For a multidimensional array:

```
b = np.array([[0,1,2,3],[4,5,6,7],[8,9,10,11]])
i = np.array([0,1,2,2]) # array of first indices
j = np.array([1,0,3,1]) # array of second indices
b[i,j] # select the elements of indices ij
```

Finally, it is possible to index an array by an array of Booleans:

```
c = np.array([[0,1,2,3],[4,5,6,7],[8,9,10,11]])
cond = (c >= 5) # array of Booleans defined by True if >= 5
 ↵   and False otherwise
c[cond] = 5 # assign the value 5 to all entries greater than
 ↵   5
```

For the following, we consider the numbers:

```
[0.9602, -0.99, 0.2837, 0.9602, 0.7539, -0.1455, -0.99,
 ↵   -0.9111, 0.9602, -0.1455, -0.99, 0.5403, -0.99, 0.9602,
 ↵   0.2837, -0.99, 0.2837, 0.9602]
```

as the results of a measurement made every 0.1 second at times between 2 and 3.7 seconds.

**a.** Since the measurements are supposed to be positive, change the data to 0 when the values are negative.

**b.** Calculate the times for which the previous measurements are maximum.

**c.** For each maximum measure, return the previous measure, the maximum measure, and the next measure. If the previous or the next measure does not exist, replace them with np.nan.

# SOLUTIONS

## SOLUTION 3.1   INTRODUCTION TO NUMPY

**a.** The `arange` function is analogous to the `range` function but returns a NumPy array:

```
np.arange(5,10) # vector (5,6,7,8,9)
np.arange(3,10,2) # vector (3,5,7,9)
```

**b.** The function `linspace` allows to generate a number of points in an interval:

```
np.linspace(2,5,10)
```

**c.** The transformations are successively given by:

```
mat0 = np.arange(1,7)
mat1 = mat0.reshape(3,2)
mat2 = mat1.reshape(2,3)
mat3 = mat2.T
```

It is possible to go directly from the first to the last matrix:

```
mat0.reshape(3,2,order='F')
```

## SOLUTION 3.2   OPERATIONS ON ARRAYS

**a.** The definition for lists:

```
def diff_list(lst):
    out = []
    for i in range(len(lst)-1):
        out.append(lst[i+1]-lst[i])
    return out
```

and the one for NumPy tables using indexing:

```
def diff_np(vec):
    return vec[1:] - vec [:-1]
```

**b.** For the lists:

```
%%timeit
result = diff_list(a_list)
```

is executed in 197 μs ± 5.48 μs. For NumPy arrays:

```
%%timeit
result = diff_np(a_np)
```

is executed in 2.07 µs ± 42.9 ns.

## SOLUTION 3.3   VANDERMONDE MATRIX

**a.** It is a double loop to fill the matrix element by element:

```
def vdm_1(x,n):
    p = np.size(x)
    vdm = np.empty((p,n+1))
    for i in range(p):
        for j in range(n+1):
            vdm[i,j] = x[i]**j
    return vdm
```

**b.** The $k$-th column of $V(x, n)$ is the vector $x$ whose elements are raised to the power $p$, so it is possible to fill the matrix column by column:

```
def vdm_2(x,n):
    p = np.size(x)
    vdm = np.empty((p,n+1))
    for k in range(n+1):
        vdm[:,k] = x**k
    return vdm
```

**c.** In order to avoid raising the vector $x$ to the power $p$, it is possible to take the previous column:

```
def vdm_3(x,n):
    p = np.size(x)
    vdm = np.ones((p,n+1))
    for j in range(1,n+1):
        vdm[:,j] = vdm[:,j-1]*x
    return vdm
```

**d.** The execution times can be determined easily with `%%timeit`:

```
n, p = 150, 100
x = np.random.rand(p)
```

```
%%timeit
vdm_1(x,n)
```

```
%%timeit
vdm_2(x,n)
```

```
%%timeit
vdm_3(x,n)
```

Not too surprisingly, the execution times are decreasing, as a double loop is slower than a single loop, and a product faster to perform than a power.

## SOLUTION 3.4  ARRAY INDEXING (!)

**a.** The easiest way is to use indexing:

```
# array of time and measures
time = np.arange(2,3.8,0.1)
data = np.cos([6,3,5,6,7,8,3,9,6,8,3,1,3,6,5,3,5,6]).round(4)
# correct negative values by putting zero instead
data[data < 0] = 0
```

**b.** The indexing allows to easily return the times of the maximum measurements:

```
# times indexed by the Booleans array of maximum measurements
time[data == data.max()]
```

**c.** To avoid having to manage side effects too manually, the modulo is useful to select elements that make sense, even if you have to replace them later:

```
# number of measures
nb = len(time)
# indices of maximum measures
ind = np.arange(nb)[data == np.max(data)]
# indices of preceding and following maximum measuresd
tab = np.array([ind-1,ind, ind+1])
# array of adjacents measures (the modulo allows us to assign
↳    a value even outside of the range of measures)
adjacents = data[tab % nb]
# modify the values outside of the range
adjacents[tab < 0] = np.nan
adjacents[tab >= nb] = np.nan
```

# CHAPTER 4

# Plotting

Plotting results from numerical or symbolic calculations is essential to analyze or interpret them. The Matplotlib module allows making very varied visualizations. The documentation of Matplotlib is available at the address: `https://matplotlib.org/users/`.

To use it, it is usual to import it like this:

```
import matplotlib.pyplot as plt
```

and often NumPy is also very useful:

```
import numpy as np
```

## Concepts covered

- one-dimensional plots

- two- and three-dimensional plots

- cobweb diagram

- bifurcation diagram

- optimization by parallelization

DOI: 10.1201/9781003565451-4

# EXERCISES

---

## EXERCISE 4.1   PLOTS

For example, the function plot can be used to represent the function $x^2$:

```
x = np.linspace(0,1,50)
y = x**2
plt.plot(x,y)
plt.show()
```

In order to define a nice figure that can be exported as in Figure 4.1, the syntax is as follows:

```
plt.figure(figsize=(8,5)) # size of the figure (in inches)
plt.title(r'Plot of $x^2$') # title of the figure (LaTeX code
    ↳   can be included)
plt.xlabel(r'$x$') # label of the horizontal axis
plt.ylabel(r'$y$') # label of the vertical axis
plt.plot(x, y, marker='o', label=r"$x^2$") # legend
plt.legend() # display the legend
plt.savefig("test.pdf") # export the figure to PDF
plt.savefig("test.png", dpi=100) # export the figure to PNG
plt.show()
```

**a.** Plot the functions $\sin(kx)$ and $\cos(kx)$ for $k = 1, 2, 3$ on $x \in [0, 2\pi]$ in the same figure. Make the graduations on the horizontal axis all $\frac{\pi}{2}$, as in Figure 4.2.

*Hint: Use the* xticks *function described at the address:* https://matplotlib.org/api/_as_gen/matplotlib.pyplot.xticks.html.

**b.** Look at the help for the imshow function and use it to plot a matrix of random numbers in $[0, 1]$ of size $10 \times 10$ as in Figure 4.3.

**c.** Plot the density and the contour lines of the function $f(x, y) = \frac{-y}{5} + e^{-x^2 - y^2}$ for $x \in [-3, 3]$ and $y \in [-3, 3]$ as in Figure 4.4.

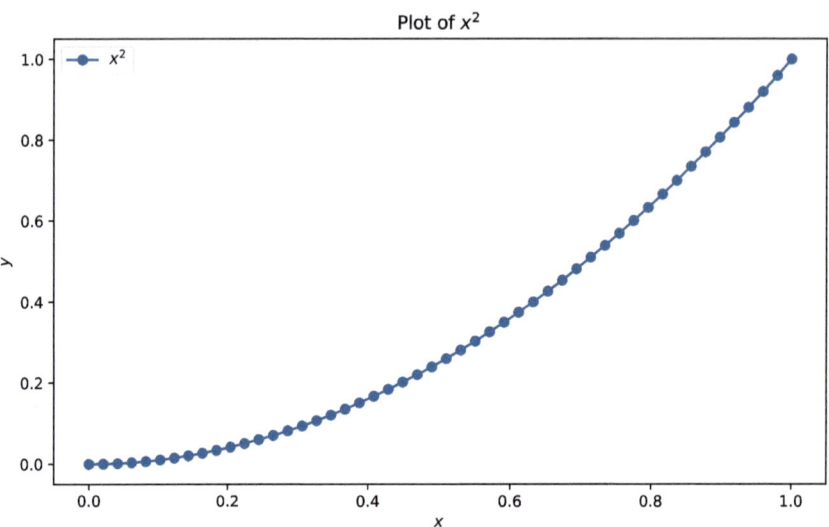

Figure 4.1    Plot of $x^2$.

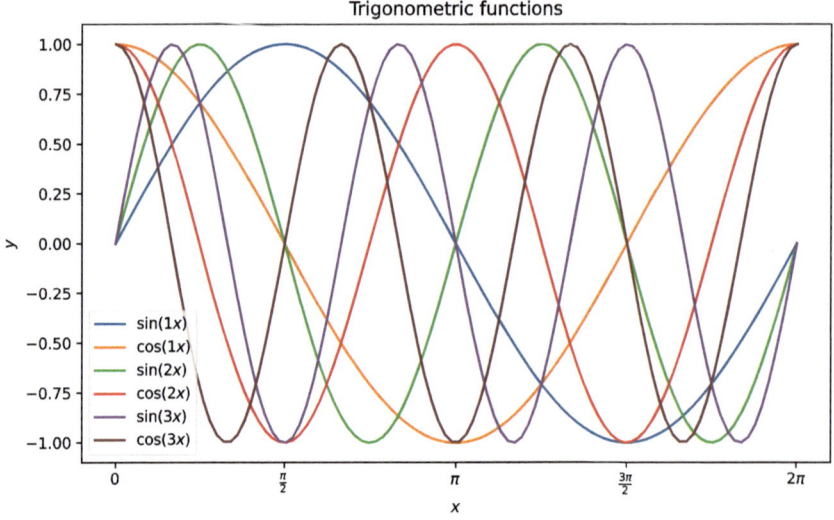

Figure 4.2    Plot of $\sin(kx)$ and $\cos(kx)$ for $k = 1, 2, 3$.

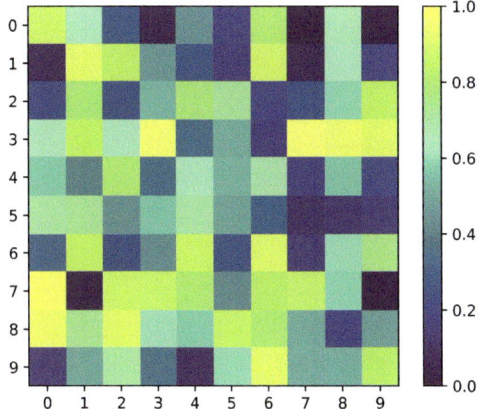

Figure 4.3    Plot of a random matrix. The colors correspond to the values of the entries of the matrix.

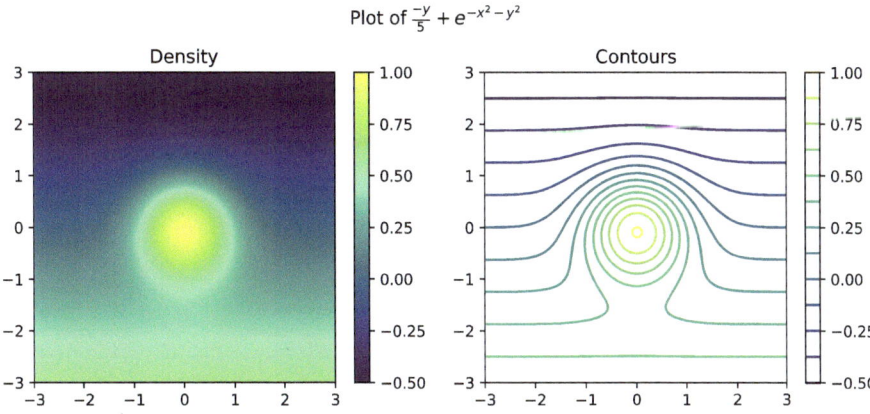

Figure 4.4    Density and contour lines of the function $f$.

## EXERCISE 4.2   DETERMINISTIC CHAOS

The goal is to study an extremely elementary model of evolution of a population. In spite of its simplistic character, this model presents many very interesting properties, in particular chaotic ones.

The proportion of a population at a discrete time $i \in \mathbb{N}$ is noted $x_i \in [0,1]$. The evolution of the population is given by $x_{i+1} = f_\mu(x_i)$, where $f_\mu : [0,1] \to [0,1]$ is the function defined by:

$$f_\mu(x) = \mu x (1-x).$$

The parameter $\mu \in [0,4]$ describes the population growth. The application $f_\mu$ is called the logistic map of parameter $\mu$.

**a.** Define the function $f_\mu$ in Python as $\mathtt{f(x, mu)} = f_\mu(x)$ and plot it for several values of $\mu$.

**b.** Why can't the value of $\mu$ be greater than 4 in the model?

**c.** Define a function that takes as argument a value of $\mu$, an initial data $x_0 \in [0,1]$, and a number $n \in \mathbb{N}$ and returns the list $S_\mu^n = (x_0, x_1, x_2, x_n)$. Modify this function so that it can take an optional parameter $m \in \mathbb{N}$ and return the list $S_\mu^{m,n} = (x_m, x_{m+1}, ..., x_n)$, i.e., remove the first $m-1$ elements from $S_\mu^n$.

**d.** Test this function by making a graphical representation of the list $S_\mu^n$ for different values of the parameters $x_0$ and $\mu$. Observe the different behaviors of the sequence.

**Cobweb diagram:**   One way to study more specifically what happens when $\mu$ varies is to make a cobweb diagram that consists in connecting by straight lines the points:

$$\{(x_0, 0), (x_0, x_1), (x_1, x_1), (x_1, x_2), (x_2, x_2), ..., (x_n, x_n), (x_n, x_{n+1})\}.$$

**e.** Define a function that returns the list of points needed to build the cobweb diagram.
*Hint: In order to use Matplotlib afterwards, it is sometimes simpler to represent a list of points $\{(x_0, y_0), (x_1, y_1), ..., (x_n, y_n)\}$ as the vector of x-coordinates and the vector of y-coordinates, i.e., $(x_0, x_1, ..., x_n)$ and $(y_0, y_1, ..., y_n)$.*

**f.** Define a function that draws the graph of the function $f_\mu$, the graph of the identity function, as well as the segments connecting the points of the previous list.

**g.** By experimenting, study qualitatively the effects of the parameters and $x_0$. Describe the behavior observed when $\mu$ increases.

**Bifurcation diagram:**   The previous experiments suggest that the behavior of the sequence $x_i$ in large time (*i.e.*, when $i$ is large) is independent of the choice of the initial condition $x_0$ but depends a lot on the value of the parameter $\mu$. The aim of this section is to represent graphically for each value of $\mu$ the set of points $S_\mu^{m,n} = (x_m, x_{m+1}, ..., x_n)$, i.e., by putting $\mu$ on the horizontal axis and all the values of

$S_\mu^{m,n}$ on the vertical axis. A good choice of parameters to clean up the diagram and keep only the long-time behavior of the system is $n = 200$ and $m = 100$, for example.

**h.** Define a function that for a given list $L$ of values of $\mu$, an initial data $x_0 \in [0,1]$ and integers $m, n \in \mathbb{N}$ returns the list of points $\{(\mu, x) : x \in S_\mu^{m,n}$ for $\mu \in L\}$.

**i.** Define a function that plot this list of points. It is suggested to take for $L$ a list of $1\,000$ values in the interval $[0,4]$.
*Hint: Use the* scatter *function of Matplotlib with* s=1 *and* edgecolor='none'.

**j.** Interpret the obtained diagram, in particular what it says about the long-time behavior of the system. Determine approximately for which values of $\mu$ the system :

- has zero as its only fixed point;

- has a unique nonzero fixed point;

- oscillates between two distinct values (cycle of length two);

- oscillates between four distinct values (cycle of length four);

- oscillates between three distinct values (cycle of length three).

**Representation of the attractor:** For values of $\mu$ close to 4, the values of the population $x_i$ seem to be more or less random. However, the system is purely deterministic in the sense that for a given initial value $x_0$, the population $x_i$ is defined without randomness. This *a priori* random behavior is called deterministic chaos. In order to notice that the points $x_i$ are not randomly determined, the goal is to represent graphically the points $(x_n, x_{n+1})$ to see that $x_{n+1}$ is not random at all with respect to $x_{n+1}$.

**k.** For each given value of $\mu$, define a function that returns the list of points:

$$\{(x_m, x_{m+1}), (x_{m+1}, x_{m+2}), \dots, (x_n, x_{n+1})\},$$

**l.** Plot these points for different values of $\mu$. For example, $n = 5\,000$ and $m = 100$ is a good choice of parameters.

**m.** How would the previous plot look like if each $x_i$ were drawn randomly in the interval $[0,1]$ independently of $x_{i-1}$?

## EXERCISE 4.3   MANDELBROT SET

The Mandelbrot set is defined as the set of points $c \in \mathbb{C}$ for which the sequence of complex numbers defined recursively by $z_0 = 0$ and

$$z_{n+1} = z_n^2 + c,$$

is bounded. It is possible to show that $c \in \mathbb{C}$ is in the Mandelbrot set if and only if $|z_n| \leq 2$ for any integer $n$.

**a.** Write a function mandelbrot(c) that checks if the point $c \in \mathbb{C}$ is in the Mandelbrot set approximately by testing the first hundred iterations.

**b.** Test the previous function with $c = 0$ and $c = 1 + i$. What is expected theoretically?

*Hint: For example, the complex number $2 + 3i$ is defined in Python as* 2+3j.

**c.** Write a function mandelbrot_set(N) that generates an array of size $N$ representing the set $c \in \{x+iy : x \in [-2, 2] \text{ and } y \in [-2, 2]\}$ and that returns an array of Booleans of size $N$ determining if the associated point is in the Mandelbrot set or not.

**d.** Using the previous function with $N = 100$, plot with imshow an approximation of the set of points belonging to the Mandelbrot set.

**e.** Adapting the previous functions, plot the logarithm of the number of iterations required before $|z_n| \leq 2$ is no longer satisfied instead of a Boolean, as represented on Figure 4.5.

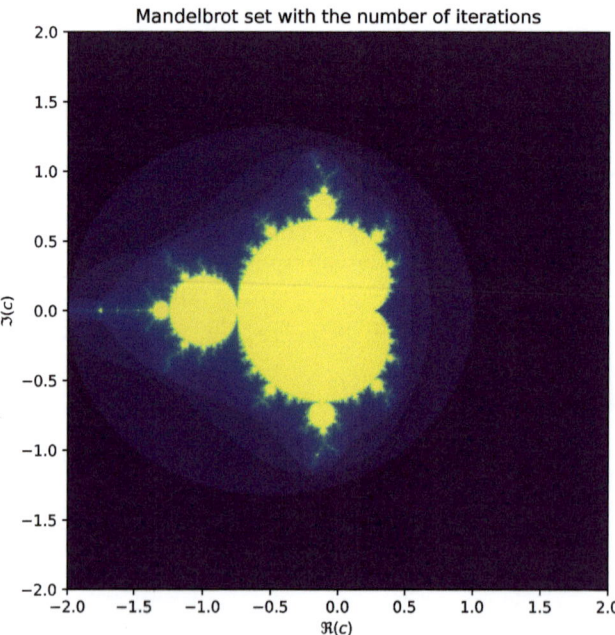

Figure 4.5 Graphical representation of the Mandelbrot set, where the color represents the logarithm of the number of iterations needed to get out of the disk of diameter two.

**f.** ‼ The previous method has the disadvantage of computing each value of $c$ sequentially, which makes the evaluation rather slow. Propose a new implementation allowing to compute in parallel all the values using NumPy indexing.

## EXERCISE 4.4   ADVANCED GRAPHICS (!)

The purpose of this exercise is to discover a range of possibilities offered by Matplotlib.

**a.** Draw the stream lines of the vector field of the Van der Pol oscillator:

$$\begin{pmatrix} y \\ -x + \mu(1 - x^2)y \end{pmatrix}$$

for different values of $\mu \in \mathbb{R}$.

**b.** Plot the parametric curve:

$$\begin{pmatrix} (1 + t^2)\sin(2\pi t) \\ (1 + t^2)\cos(2\pi t) \\ t \end{pmatrix}$$

for $t \in [-2, 2]$.

**c.** Plot the function of two variables:

$$f(x, y) = \sin\left(\sqrt{x^2 + y^2}\right)$$

in three dimensions for $x \in [-5.5]$ and $y \in [-5.5]$.

**d.** !! Represent the given Möbius strip as a parametric surface:

$$\begin{pmatrix} (3 + v\cos(u/2))\cos u \\ (3 + v\cos(u/2))\sin u \\ v\sin(u/2) \end{pmatrix}$$

for $u \in [0, 2\pi]$ and $v \in [-1, 1]$.

**e.** !! Look at the examples available at the address: `https://matplotlib.org/tut orials/introductory/sample_plots.html` and choose two to understand and modify.

# SOLUTIONS

---

## SOLUTION 4.1   PLOTS

**a.** To limit the repetition of code, it is wise to make a loop to obtain Figure 4.1:

```python
x = np.linspace(0,2*np.pi,100)
plt.figure(figsize=(8,5))
plt.title(r'Trigonometric functions')
plt.xlabel(r'$x$')
plt.ylabel(r'$y$')
plt.xticks([0,np.pi/2, np.pi, 3*np.pi/2,2*np.pi],
    [r'$0$',r'$\frac{\pi}{2}$', r'$\pi$',
    r'$\frac{3\pi}{2}$', r'$2\pi$'])
for k in range(1,4):
    plt.plot(x, np.sin(k*x), label=r"$\sin({}x)$".format(k))
    plt.plot(x, np.cos(k*x), label=r"$\cos({}x)$".format(k))
plt.legend()
```

**b.** The `vmin` and `vmax` parameters of `imshow` are used to specify the range of the legend bar:

```python
A = np.random.rand(10,10)
plt.figure(figsize=(5,4))
plt.imshow(A, vmin=0, vmax=1)
plt.colorbar()
plt.xticks(range(A.shape[0]))
plt.yticks(range(A.shape[1]))
plt.show()
```

**c.**  The function `meshgrid` allows to transform two one-dimensional arrays representing coordinates into two two-dimensional arrays representing functions of two variables $(x,y) \mapsto x$ and $(x,y) \mapsto y$. It is then sufficient to combine these two-dimensional arrays:

```python
# define 100 points in [-3,3] for x and y
x = y = np.linspace(-3.0, 3.0, 100)
# construct the matrices representing X and Y respectively
X, Y = np.meshgrid(x, y)
# construct the matrix representing the function
Z = -Y/5 + np.exp(-X**2 - Y**2)

# create a figure with a title
fig = plt.figure(figsize=(12,5))
fig.suptitle(r'Plot of $\frac{-y}{5} + e^{-x^2-y^2}$')
```

```
# create a sub-figure for the density
sub = fig.add_subplot(1,2,1)
sub.set_title('Density')
# density plot
im = sub.imshow(Z, interpolation='bilinear', origin='lower',
  ↪  extent=[-3,3,-3,3], vmin=-0.5, vmax=1)
# legend with 7 ticks
plt.colorbar(im, ax=sub, ticks=np.linspace(-0.5,1,7))

# create a new figure of the contours
sub = fig.add_subplot(1,2,2)
sub.set_title('Contours')
# contour plot with 13 levels
im = sub.contour(X, Y, Z, levels=np.linspace(-0.5,1,13))
# legend with 7 ticks
plt.colorbar(im, ax=sub, ticks=np.linspace(-0.5,1,7))
# force same scale on vertical and horizontal axes
sub.set_aspect('equal')

# reduces the size of the margins
fig.tight_layout()
fig.subplots_adjust(top=0.9)
plt.show()
```

## SOLUTION 4.2   DETERMINISTIC CHAOS

**a.** We need to define the function $f_\mu$:

```
def f(x,mu):
    return mu*x*(1-x)
```

then define a function to plot it:

```
def plot_f(mu):
    x = np.linspace(0,1,100)
    fx = f(x,mu)
    plt.plot(x,fx)
```

To test:

```
plot_f(2)
```

**b.** By taking $\mu \geq 4$, the function $f_\mu$ takes values greater than 1 and thus leaves the domain of definition for the next iteration:

```
plot_f(5)
```

**c.** The idea is to define an empty vector, then to fill it element by element, finally to select the requested part:

```python
def simulate(mu, x0, n, m=0):
    vec = np.zeros(n+1)
    vec[0] = x0
    for i in range(n):
        vec[i+1] = f(vec[i], mu)
    return vec[m:]
```

**d.** The easiest way is to plot the curves obtained for different values of $\mu$:

```python
plt.figure(figsize=(8,5))
plt.title(r"Evolution of $x_{i+1}=f_\mu(x_i)$")
plt.xlabel("$i$")
plt.ylabel("$x_i$")
for mu in (0.5,1,1.5,2,2.5,3,3.5,3.9,3.90001):
    data = simulate(mu,0.1,50)
    plt.plot(data, label=r"$\mu={}$".format(mu))
plt.legend()
```

Figure 4.6 allows to observe that for values of $\mu$ lower than about 1, the sequence stabilizes around 0. For values of $\mu$ between 1 and 3, the sequence seems to stabilize around a non-zero value. When $\mu$ is even larger, the sequence oscillates.

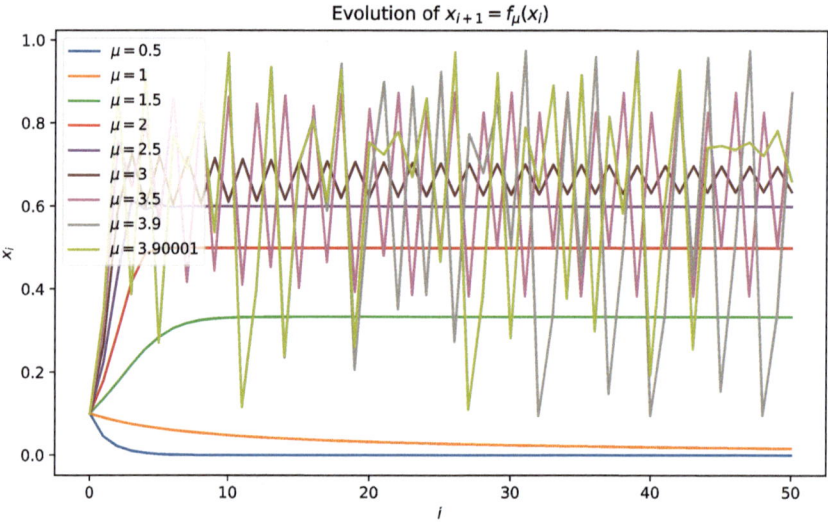

**Figure 4.6** Evolution of the logistic sequence for different values of $\mu$ starting from $x_0 = 0.1$.

**e.** A quick and elegant way to build this list of points is to use the `repeat` function of NumPy:

```
def cobweb(data):
    x = np.repeat(data,2)[:-1]
    y = np.repeat(data,2)[1:]
    y[0]=0
    return (x,y)
```

**f.** The previously defined functions must be combined to obtain the cobweb diagram:

```
def plot_cobweb(mu,x0,n):
    plt.title(f"Cobweb diagram for $\\mu={mu}$")
    # plot f_mu
    x = np.linspace(0,1,100)
    fx = f(x,mu)
    plt.plot(x,fx)
    # plot identity
    plt.plot([0,1],[0,1])
    # plot segments
    data = simulate(mu,x0,n)
    segments = cobweb(data)
    plt.plot(segments[0],segments[1])
```

**g.** We choose different values of $\mu$ to obtain the cobweb diagrams of Figure 4.7:

```
fig = plt.figure(figsize=(8,8))
for i,mu in enumerate([2.5,3.2,3.5,3.8]):
    plt.subplot(2,2,i+1)
    plot_cobweb(mu,0.1,100)
```

For $0 < \mu < 3$, the sequence converges to one of the two fixed points of $f_\mu$, i.e., a solution of $f_\mu(x) = x$. For $3 < \mu < 3.45$, the sequence oscillates between two values and then it seems to oscillate between several values or in a very random way.

**h.** The function full allows to build a constant vector equal to mu in this case:

```
def simulate_range(L,x0=0.5,n=200,m=100):
    all_x = []
    all_y = []
    for mu in L:
        y = simulate(mu,x0,n,m=m)
        x = np.full(len(y),mu)
        all_x.append(x)
        all_y.append(y)
    return(np.array(all_x),np.array(all_y))
```

**i.** The scatter function is used to represent a cloud of points. The options s=1 and edgecolor='none' allow to reduce the size of the points so that they overlap less. The following code allows to obtain the bifurcation graph of Figure 4.8:

```python
def plot_bif(points,xlim=(0,4),ylim=(0,1), **options):
    plt.figure(figsize=(8,5))
    plt.title("Bifurcation diagram")
    plt.xlabel(r"$\mu$")
    plt.scatter(points[0], points[1], c=points[0], s=1,
      ↳ edgecolor='none', cmap='jet', **options)
    plt.xlim(xlim)
    plt.ylim(ylim)
```

To test:

```python
points = simulate_range(np.linspace(0,4,1000))
plot_bif(points, xlim=(0,4))
```

**j.** When $0 < \mu < 1$, the sequence converges to the zero fixed point. When $1 < \mu < 3$, the sequence converges to the unique non-zero fixed point. For $3 < \mu < 1+\sqrt{6}$, the sequence oscillates between two distinct values. For $1 + \sqrt{6} < \mu < 3.54409$, the sequence oscillates between four distinct values. Finally, the smallest value of $\mu$ with an oscillation between three values is for $\mu = 1 + \sqrt{8}$.

**k.** The following function is intended to act on the output of the `simulate_range` function. We choose to return also the corresponding value of $\mu$:

```python
def attr(points):
    data = points[1]
    x = data[:,:-1]
    y = data[:,1:]
    c = points[0][:,:-1]
    return (x,y,c)
```

**l.** The points returned by `attr` are colored according to the value of $\mu$, which allows to represent the attractor of Figure 4.9:

```python
data = simulate_range(np.linspace(3.5,4,10), x0=0.5, n=5000,
  ↳ m=100)
pts = attr(data)
c = np.linspace(0,4,100)
plt.figure(figsize=(8,5))
plt.title("Plot of the attractor")
plt.xlabel("$x_n$")
plt.ylabel("$x_{n+1}$")
plt.xlim(0,1); plt.ylim(0,1)
plt.scatter(pts[0], pts[1], c=pts[2], s=2, edgecolor='none')
plt.colorbar()
```

For a given value of $\mu$, the points are as expected, located on the graph of the function $f_\mu$. The larger the value of $\mu$ is, the larger the attractor, *i.e.*, the set of limit points, is.

**m.** The points would not be aligned on the graph of the function $f_\mu$.

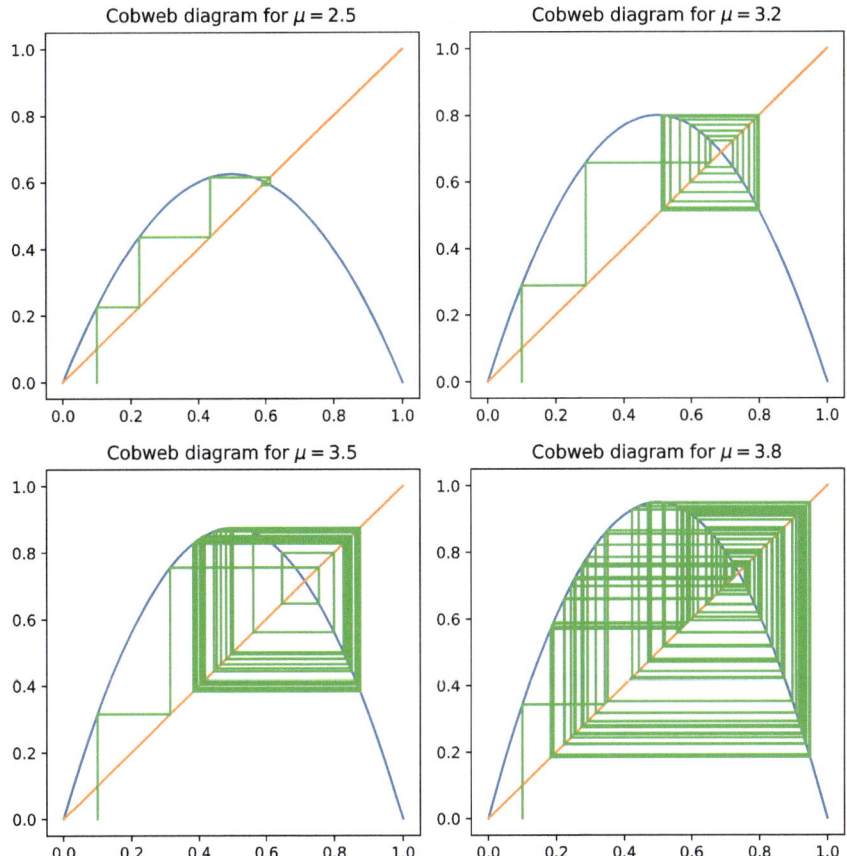

Figure 4.7　At $\mu = 2.5$, the sequence converges to a fixed point, then for $\mu = 3.2$, it oscillates between two values. For the last two values of $\mu$, the sequence seems to oscillate between several values or even randomly.

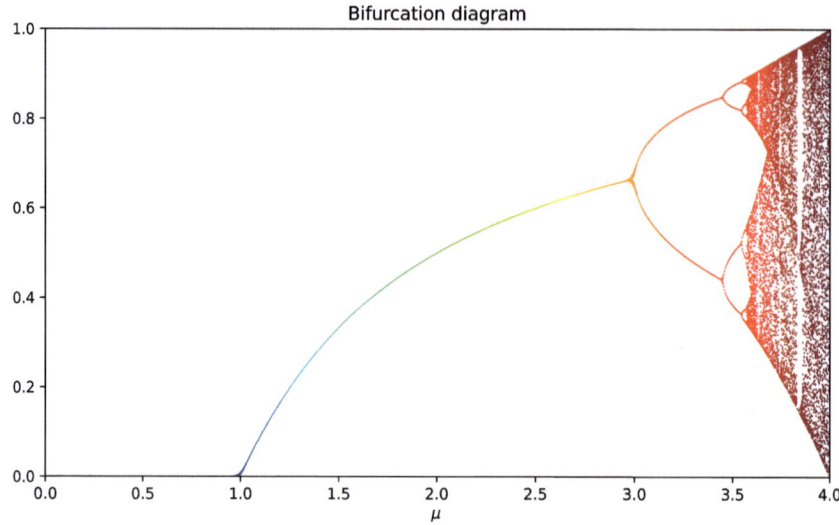

**Figure 4.8**    Bifurcation diagram for the logistic map depending on $\mu$.

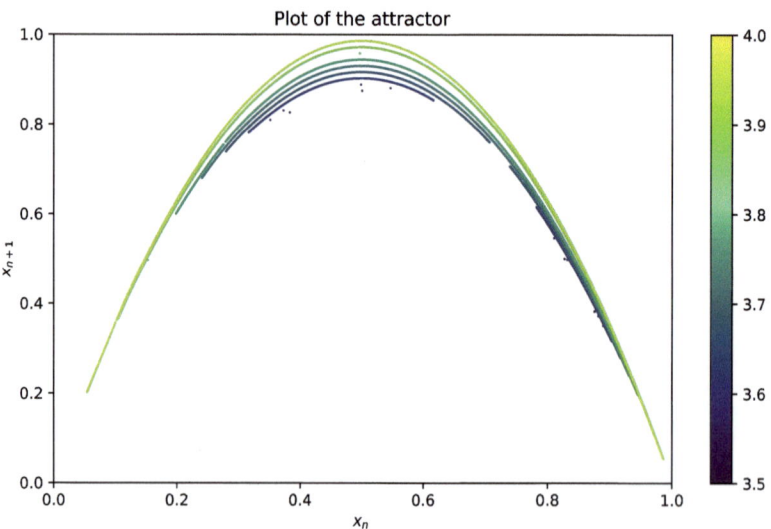

**Figure 4.9**    Plot of the attractor of the logistic map for different values of $\mu$.

## SOLUTION 4.3   MANDELBROT SET

**a.** We have to iterate the recurrence relation and check each time if the condition $|z_n| \le 2$ is satisfied:

```python
def mandelbrot(c,max=100):
    z = 0
    for i in range(max):
        z = z**2+c
        if abs(z)>2:
            return False
    return True
```

**b.** The point $c = 0$ is clearly in the Mandelbrot set since $z_n = 0$ for all $n$:

```python
mandelbrot(0)
```

For $c = 1 + i$, then $z_1 = 1 + i$ and $z_2 = 1 + 3i$ which is of modulus greater than 2, so $c$ is not in the Mandelbrot set:

```python
mandelbrot(1+1j)
```

**c.** The first step is to generate the grid; the second is to build the requested table:

```python
def mandelbrot_set(N):
    # construct the grid
    lst = np.linspace(-2,2,N)
    x, y = np.meshgrid(lst,lst)
    c = x + 1j*y
    # define the result array with adapted type
    out = np.zeros_like(c, dtype=type(mandelbrot(0)))
    # fill the result array
    for i in range(N):
        for j in range(N):
            out[i,j] = mandelbrot(c[i,j])
    return out
```

**d.** This represents the previously constructed array to get Figure 4.10:

```python
plt.figure(figsize=(6,6))
plt.title("Mandelbrot set")
plt.xlabel(r"$\Re(c)$")
plt.ylabel(r"$\Im(c)$")
plt.imshow(mandelbrot_set(100), extent=[-2,2,-2,2])
```

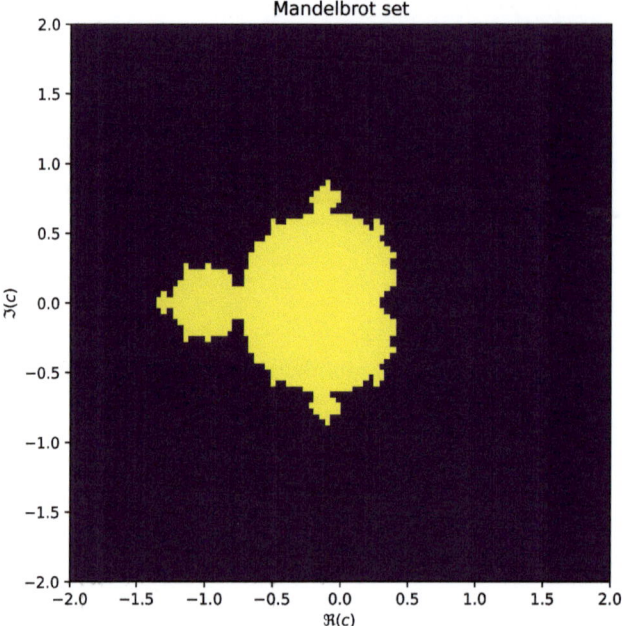

Figure 4.10   Graphical representation of an approximation of the Mandelbrot set.

**e.** Only a slight modification of the mandelbrot function is required:

```python
def mandelbrot(c,max=100):
    z = 0
    for i in range(1,max+1):
        z = z**2+c
        if abs(z)>2:
            return np.log(i)
    return np.log(max)
```

to draw the same thing:

```python
plt.figure(figsize=(6,6))
plt.title("Mandelbrot set with number of iterations")
plt.xlabel(r"$\Re(c)$")
plt.ylabel(r"$\Im(c)$")
plt.imshow(mandelbrot_set(200), extent=[-2,2,-2,2])
```

**f.** The idea is to iterate the whole array corresponding to the grid at the same time, except for the values that have already diverged:

```python
def mandelbrot_plot(N, max=1000):
    # construct the grid
    lst = np.linspace(-2,2,N)
    x, y = np.meshgrid(lst,lst)
    c = x + 1j*y

    # output array
    out = np.zeros_like(c, dtype=float)
    # current iteration
    z = np.zeros_like(c, dtype=complex)
    # Boolean array of the z that have not diverged
    cond = np.abs(z) <= 2
    # iterations
    for i in range(1,max+1):
        # iteration on the z that have not diverged
        z[cond] = z[cond]**2 + c[cond]
        # update the condition
        cond[cond] = np.abs(z[cond]) <= 2
        # add the logarithm for the z that diverged at this
        ↳  iteration
        div = np.logical_and(~cond,out==0)
        out[div] = np.log(i)
    # add the logarithm for the remaining z
    out[cond] = np.log(max)

    # plot
    plt.figure(figsize=(6,6))
    plt.title("Mandelbrot set with the number of iterations")
    plt.xlabel(r"$\Re(c)$")
    plt.ylabel(r"$\Im(c)$")
    plt.imshow(out, extent=[-2,2,-2,2])
```

Figure 4.5 represents the evaluation of this function for N=1000.

## SOLUTION 4.4   ADVANCED GRAPHICS (!)

**a.** The streamplot function allows to draw the streamlines as in Figure 4.11:

```python
fig = plt.figure(figsize=(8,8))
fig.suptitle("Streamlines for the Van der Pol oscillator")
# discretization in space
l = np.linspace(-3,3,100)
X, Y = np.meshgrid(l,l)

for i,mu in enumerate([-1,0,1,5]):
    sub = fig.add_subplot(2,2,i+1)
```

```
        sub.set_title(f"$\\mu = {mu}$")

        # definition of the vector field
        U = Y
        V = -X + mu*(1-X**2)*Y
        # definition of the color
        C = np.sqrt(U**2+V**2) # color

        # streamlines
        plt.streamplot(X, Y, U, V, color=C)
        sub.set_xlim([-3,3])
        sub.set_ylim([-3,3])
        sub.set_aspect('equal')
fig.tight_layout()
fig.subplots_adjust(top=0.91)
```

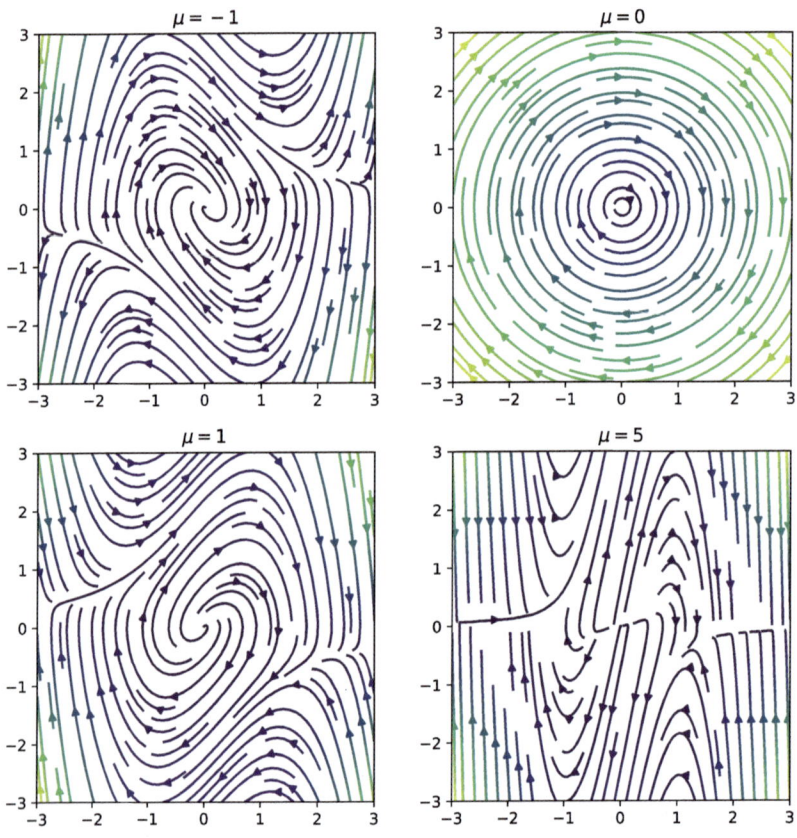

Figure 4.11    Streamlines of the Van der Pol oscillator.

**b.** It is a matter of initializing a 3d graphic and then using the `plot` function with three arguments to get Figure 4.12:

```
fig = plt.figure(figsize=(6,5.5))
# initialization of a 3d plot
ax = fig.add_subplot(111, projection='3d')
ax.set_title("Parametric curve", y=1.02)
ax.set_xlabel("$x$")
ax.set_ylabel("$y$")
ax.set_zlabel("$z$")

# discretization of the parameter
t = np.linspace(-2, 2, 200, endpoint=True)
# definition of the points corresponding to the curve
r = t**2 + 1
x = r*np.sin(2*np.pi*t)
y = r*np.cos(2*np.pi*t)
z = t

# plot the parametric curve
ax.plot(x, y, z)
```

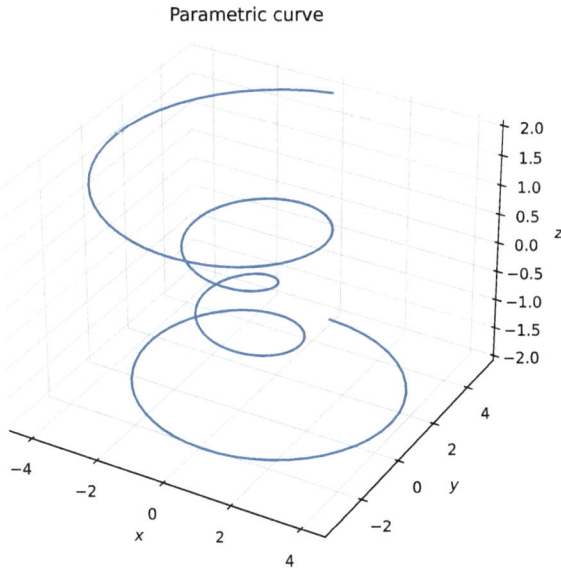

Figure 4.12  Plot of a parametric curve in three-dimensional space.

**c.** The function `plot_surface` allows to represent a function of two variables in three dimensions as in Figure 4.13:

```python
fig = plt.figure(figsize=(7,5.8))
# initialization of a 3d plot
ax = fig.add_subplot(111, projection='3d')
ax.set_title("Plot of the function $f(x,y)$", y=1.02)
ax.set_xlabel("$x$")
ax.set_ylabel("$y$")
# construction of the coordiantes
xy = np.linspace(-5, 5, 100, endpoint=True)
X, Y = np.meshgrid(xy,xy)
# construction of the function
R = np.sqrt(X**2 + Y**2)
Z = np.sin(R)
# plot the surface with data between -1 and 1
surf = ax.plot_surface(X, Y, Z, vmin=-1, vmax=1,
    cmap="coolwarm")
# range of vertical axis between -1 and 1
ax.set_zlim(-1, 1)
# legend bar
fig.colorbar(surf)
```

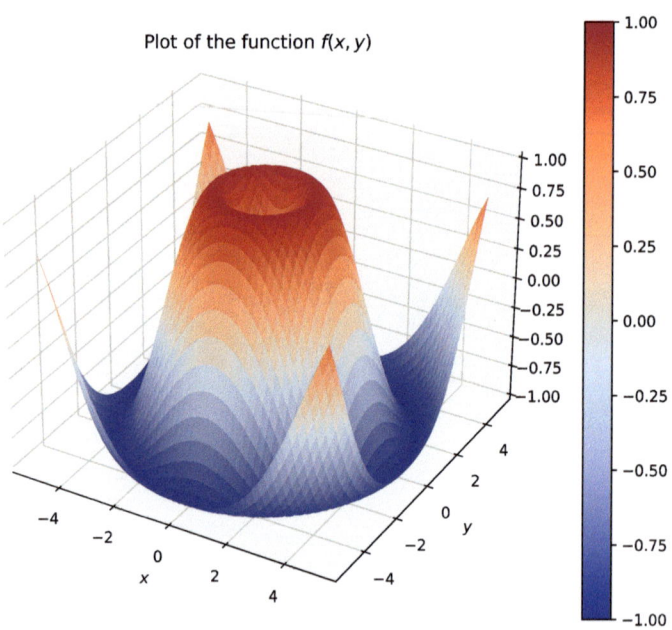

Figure 4.13  Three-dimensional representation of a function of two variables.

**d.** To represent a parametric surface, one must first define the parameter space and then triangulate it before defining the surface itself to produce Figure 4.14:

```python
# import triangulation module
import matplotlib.tri as mtri

fig = plt.figure(figsize=(6.5,6))
ax = fig.add_subplot(111, projection='3d')
ax.set_title("Parametric surface: the Möbius strip", y=1.02)
ax.set_xlabel("$x$")
ax.set_ylabel("$y$")
ax.set_zlabel("$z$")

# meshing of the parameters
u = np.linspace(0, 2*np.pi, 50, endpoint=True)
v = np.linspace(-1, 1, 10, endpoint=True)
u, v = np.meshgrid(u, v)
u, v = u.flatten(), v.flatten()

# definition of the Möbius strip
x = (3 + v*np.cos(u/2))*np.cos(u)
y = (3 + v*np.cos(u/2))*np.sin(u)
z = v*np.sin(u/2)

# triangulation of the parameters space
tri = mtri.Triangulation(u, v)

# plot the strip and adjust the axes
ax.plot_trisurf(x, y, z, triangles=tri.triangles,
    ↳  cmap="viridis")
ax.set_xlim(-3.2,3.2)
ax.set_ylim(-3.2,3.2)
ax.set_zlim(-2,2)
```

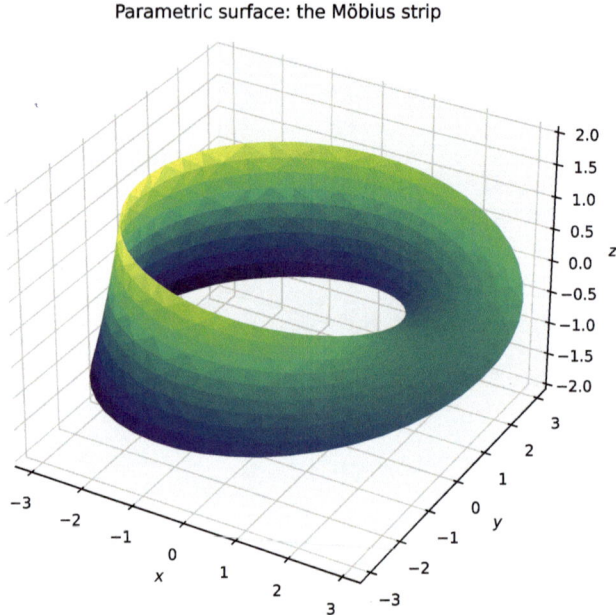

Parametric surface: the Möbius strip

Figure 4.14    Representation of the Möbius strip as a parametric surface.

# CHAPTER 5

# Integration

The goal is to obtain an approximation of a definite integral of the type:

$$J = \int_a^b f(x)\,\mathrm{d}x$$

for some function $f : [a, b] \to \mathbb{R}$ too complicated to *a priori* determine the value of $J$ by hand. Deterministic and probabilistic approximation methods will be introduced to obtain an approximation $\tilde{J}$ of $J$.

## Concepts covered

- classical methods (rectangles, trapezoids, and Simpson)

- Monte Carlo method

- convergence speed

- statistics

DOI: 10.1201/9781003565451-5

# EXERCISES

---

## EXERCISE 5.1   RECTANGLE RULE

The rectangle rule is based on the definition of the integral in the Riemann sense. The first step is to split the interval $[a, b]$ into $N$ intervals $[x_n, x_{n+1}]$ of the same size $\delta = \frac{b-a}{N}$, i.e., $x_n = a + n\delta$ for $n \in \{0, 1, \dots, N-1\}$. The second step consists in assuming that the function $f$ is constant on each interval $[x_n, x_{n+1}]$, thus to make the approximation:

$$J_n = \int_{x_n}^{x_{n+1}} f(x)\,dx \approx \delta f(\tilde{x}_n),$$

for $\tilde{x}_n$ a certain value to choose in the interval $[x_n, x_{n+1}]$. The choice of $\tilde{x}_n$ can, for example, be done by $\tilde{x}_n = x_n + \alpha\delta$ with $\alpha \in [0, 1]$. Finally, the approximation of $J$ is given by the sum of the approximations of $J_n$,

$$\tilde{J} = \sum_{n=0}^{N-1} \delta f(\tilde{x}_n).$$

Assuming that $f \in C^1([a, b])$, it is then possible to show that the rectangle rule converges and that its speed of convergence is of order one. A numerical method is of order $k$ if the error between the numerical approximation and the exact result is of order $N^{-k}$.

**a.** Choose a continuous function $f : [a, b] \rightarrow \mathbb{R}$ and define the corresponding Python function `f(x)`. To test the code, it is wise to choose a function $f$ whose integral can be easily computed by hand.
*Hint: The list of basic mathematical functions available in Python in the* math *module is available at the address:* `https://docs.python.org/3/library/math.htm` *1. Note that NumPy also defines mathematical functions, see the documentation at the address:* `https://numpy.org/doc/stable/reference/routines.math.h tml`.

**b.** Write a function `rectangles(f,a,b,N)` that returns the approximation of the integral $J$ by the rectangle rule, for example, by choosing $\tilde{x}_n = x_n$, i.e., the left edge of the interval $[x_n, x_{n+1}]$.
*Hint: It is not necessary to store all the values of the approximations of $J_n$, but it is possible to increment a variable for each approximation of $J_n$.*

**c.** Modify the previous function so that it takes an optional parameter `alpha` determining the choice of parameter $\alpha \in [0, 1]$.

**d.** Write a function `plot_rectangles(f,a,b,N,alpha=0.5)` that graphically represents the approximation by the rectangle rule.

**e.** Determine empirically the speed of convergence of the rectangle rule as a function of $N$.

**f.** Determine analytically the convergence of the rectangle rule. What are the necessary assumptions on $f$?
*Hint: Use the mean value theorem.*

## EXERCISE 5.2   TRAPEZOIDAL RULE

The trapezoidal rule is based on a linear approximation on each interval $[x_n, x_{n+1}]$, more specifically:

$$J_n = \int_{x_n}^{x_{n+1}} f(x)\,dx \approx \delta \frac{f(x_n) + f(x_{n+1})}{2}.$$

**a.** Write a Python function `trapezes(f,a,b,N)` that returns the approximation of the integral $J$ by the trapezoidal rule. Test the function `trapezes(f,a,b,N)` for different functions $f$.

**b.** Is your implementation of the function `trapezes(f,a,b,N)` optimal in terms of the number of evaluations of $f$ performed compared to the number of evaluations needed? An optimal implementation of the function `trapezes(f,a,b,N)` should perform $N + 1$ evaluations of $f$.

**c.** Determine empirically the speed of convergence of the trapezoidal rule as a function of $N$.

**d.** ! Determine analytically the convergence of the trapezoidal rule. What are the necessary assumptions on $f$?

## EXERCISE 5.3   MONTE CARLO METHOD

The Monte Carlo method (named after casinos, not a person) is a probabilistic approach to approximate the value of an integral. The basic idea is that the integral $J$ can be seen as the expectation of a uniform random variable $X$ on the interval $[a, b]$:

$$J = \int_a^b f(x)\,dx = (b - a)\mathbb{E}(f(X)).$$

By the law of large numbers, this expectation can be approximated by the empirical mean:

$$\tilde{J} = \frac{b - a}{N} \sum_{i=0}^{N-1} f(x_i),$$

where $x_i$ are drawn randomly in the interval $[a, b]$ with a uniform probability distribution.

**a.** Write a function `montecarlo(f,a,b,N)` that determines an approximation $\tilde{J}$ of $J$ by the Monte Carlo method.
*Hint: To generate a vector of random numbers, the* `random` *sub-module of NumPy can be useful, see the documentation at the address:* https://numpy.org/doc/st able/reference/random/.

**b.** Modify the previous function, so that it returns in addition to the mean $\tilde{J}$ also the empirical variance:

$$\tilde{V} = \frac{(b-a)^2}{N} \sum_{i=0}^{N-1} \left( f(x_i) - \frac{\tilde{J}}{b-a} \right)^2.$$

**c.** Study empirically the convergence of the Monte Carlo method as a function of $N$ by making for each value of $N$ a statistic on $k$ executions. More precisely, this consists in making $k$ evaluations of $\tilde{J}$ by the Monte Carlo method and to calculate the mean and the variance of the $k$ results obtained.

**d.** Determine analytically the convergence of the Monte Carlo method. What are the necessary assumptions on $f$?
*Hint: Use the central limit theorem.*

## EXERCISE 5.4   SIMPSON'S RULE (!)

Simpson's rule consists in approximating the function $f$ on each interval $[x_n, x_{n+1}]$ by a polynomial of degree 2. The most natural choice is the polynomial $p_n$ of degree 2 passing through the points $(x_n, f(x_n))$, $(\frac{x_n+x_{n+1}}{2}, f(\frac{x_n+x_{n+1}}{2}))$, and $(x_{n+1}, f(x_{n+1}))$.

**a.** Determine the explicit form of the polynomial $p_n$.
*Hint: The polynomial* $L(x) = \frac{(x-c)(x-b)}{(a-c)(a-b)}$ *takes the value 1 at $x = a$ and the value 0 at $x = b$ and $x = c$. Make a linear combination of three such polynomials.*

**b.** Compute the approximation given by $J_n \approx \int_{x_n}^{x_{n+1}} p_n(x)\, dx$.
*Hint: It is possible to calculate this integral by hand or to do it with the SymPy module, see the documentation at the address:* https://docs.sympy.org/latest/mo dules/integrals/integrals.html.

**c.** Simplify by hand the sum $\tilde{J}$ of the approximations of $J_n$.
*Answer: The result is:*

$$\tilde{J} = \frac{\delta}{3} \left[ \frac{f(b)-f(a)}{2} + \sum_{n=0}^{N-1} \left( f(x_n) + 2f\left( \frac{x_n + x_{n+1}}{2} \right) \right) \right].$$

**d.** Write a function `simpson(f,a,b,N)` to approximate $J$ with Simpson's rule.

**e.** Compare the accuracy, *i.e.*, the convergence speed of the rectangle, trapezoid, and Simpson methods as a function of $N$ for a smooth function and the function $f(x) = \sqrt{1 - x^2}$ on $[0, 1]$ (whose integral is $\frac{\pi}{4}$).

**f.** If not already done, propose a parallel implementation of Simpson's rule using NumPy indexing.

## EXERCISE 5.5   INTEGRATION WITH SCIPY (!!)

The above and other integration methods are defined in the `integrate` module of SciPy. This module allows in particular to handle more complicated cases: singular, generalized, or multidimensional integrals.

**a.** Define a function `E(n,x)` computing numerically the following integral:

$$E_n(x) = \int_1^\infty \frac{e^{-xt}}{t^n} \, dt \,.$$

*Hint: Read the documentation of the SciPy* `integrate` *sub-module available at the address:* `https://docs.scipy.org/doc/scipy/tutorial/integrate.html`.

**b.** Determine an approximation of the double integral:

$$I = \int_0^\pi \left( \int_0^y x \sin(xy) \, dx \right) dy \,.$$

# SOLUTIONS

## SOLUTION 5.1   RECTANGLE RULE

**a.** We choose a function that gives 2 when integrated over $[0, 1]$:

```python
import numpy as np
def f(x):
    return 3*x**2 + 2*x - 1 + np.sin(2*np.pi*x) +
        ↪ np.exp(-x)*np.e/(np.e-1)
Jexact = 2
```

**b.** The sum is calculated iteratively:

```python
def rectangles(f,a,b,N):
    delta = (b-a)/N
    J = 0
    for i in range(N):
        x = a + delta*i
        J += f(x)*delta
    return J
```

We find the expected result:

```python
rectangles(f,0,1,100)
```

**c.** It is a question of barely modifying the previous function:

```python
def rectangles(f,a,b,N,alpha=0.5):
    delta = (b-a)/N
    J = 0
    for i in range(N):
        x = a + delta*(i+alpha)
        J += f(x)*delta
    return J
```

We find the expected result:

```python
rectangles(f,0,1,100,alpha=0.5)
```

**d.** It is convenient to use NumPy to calculate the points:

```python
import matplotlib.pyplot as plt
def plot_rectangles(f,a,b,N,alpha=0.5):
    plt.figure(figsize=(8,5))
```

```
plt.title(f"Rectangle rule for $\\alpha = {alpha}$ and $N
 ↳  = {N}$")
# vectorize f
f = np.vectorize(f)
# plot the function f
delta = (b-a)/N
x = np.linspace(a+alpha*delta, b+alpha*delta, N,
 ↳  endpoint=False)
y = f(x)
plt.plot(x,y,"bo-")
# plot the boxes
bx = np.linspace(a, b, N+1, endpoint=True)
for i in range(N):
    # abscissa and ordinate of rectangles
    x_rect = [bx[i], bx[i], bx[i+1], bx[i+1], bx[i]]
    y_rect = [0, y[i], y[i], 0, 0]
    plt.plot(x_rect, y_rect,"r")
```

The following command allows you to test and obtain Figure 5.1:

```
plot_rectangles(f,0,1,10,alpha=0.2)
```

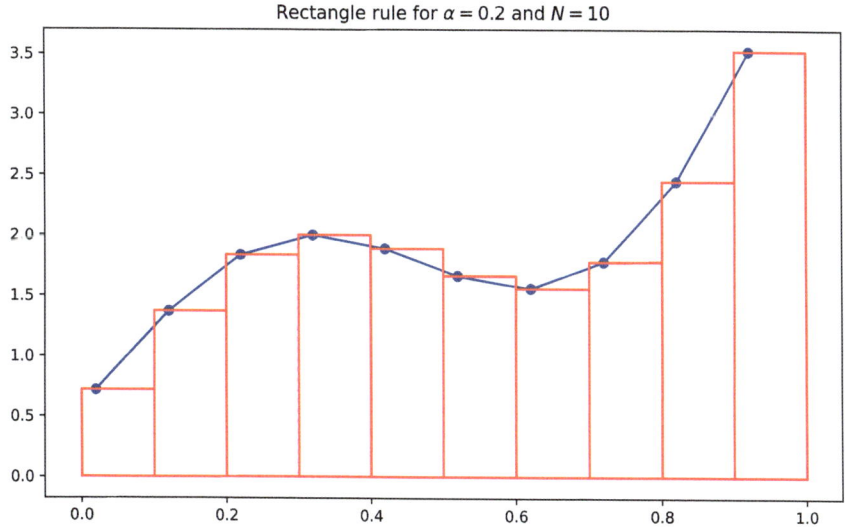

Figure 5.1   Graphical representation of the rectangle rule, with the height of the rectangles taken from the fifth of each interval.

**e.** To determine the speed of convergence, we compute the error $E_N = |\check{J} - J|$ between the exact value of the integral and the value found by making $N$ subdivisions. Then, we plot $N^k E_N$ as a function of $N$ and find the $k$ that makes the graph asymptotically constant:

```
# list of values of N
list_N = np.arange(10,1000)
# apply the function on each element of the list
data = np.vectorize(lambda N:
    ↪  rectangles(f,0,1,N,alpha=0))(list_N)
plt.figure(figsize=(8,5))
plt.title("Convergence of order one of the rectangle rule")
plt.xlabel("$N$")
plt.ylabel(r"$N \times E_N$")
plt.ylim(0,3)
plt.plot(list_N, np.abs(data-Jexact)*list_N)
```

Figure 5.2 allows to conclude that the error of the method is $E_N \propto N^{-1}$ at least for this example with $\alpha = 0$.

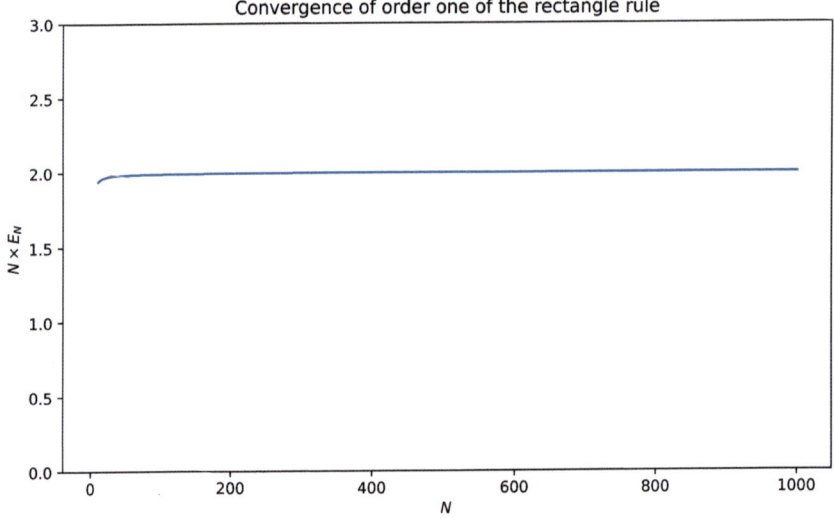

Figure 5.2    Error $E_N = |\check{J} - J|$ multiplied by $N$ as a function of $N$ allowing to conclude that the rectangle rule is of order one when $\alpha = 0$.

Another more standard way to do this is to plot $E_N$ as a function of $N$ in logarithmic scales:

```
plt.figure(figsize=(8,5))
plt.title("Order of convergence of the rectangle rule")
plt.xlabel("$N$")
for alpha in (0,0.7,0.5):
    data = np.vectorize(lambda N:
    ↪ rectangles(f,0,1,N,alpha=alpha))(list_N)
    plt.loglog(list_N, np.abs(data-Jexact), label=f"$E_N$ for
    ↪ $\\alpha={alpha}$")
for i in (1,2):
    plt.loglog(list_N, 1/list_N**i, label=f"$N^{{-{i}}}$")
plt.legend()
```

In the representation of Figure 5.3, the slope of an empirical line corresponds to the order of convergence of the method. In addition, the functions $N^{-1}$ and $N^{-2}$ are also represented, which makes it possible to compare the slopes and to deduce that the rectangle rule is of order one in general, except for $\alpha = 0.5$ where it is of order two. The case $\alpha = 0.5$ is called the mid-point rule. It would also be possible to perform linear regressions to obtain the precise values of the slopes.

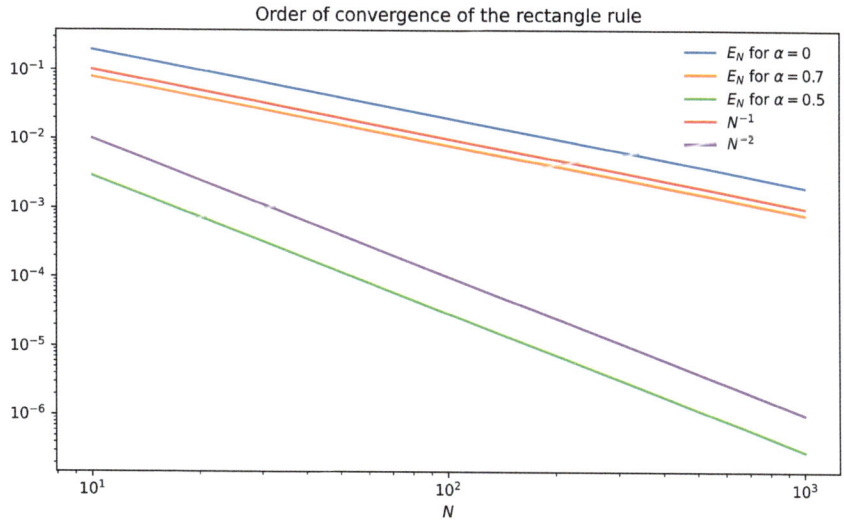

Figure 5.3 Error $E_N$ as a function of $N$ in logarithmic scales for different values of $\alpha$. The rectangle rule is therefore of order one in general, but is of order two for $\alpha = 0.5$.

**f.** For each value of $n$ and $x \in [x_n, x_{n+1}]$, by the mean value theorem, there exists $c_n$ such that:

$$f(x) - f(\tilde{x}_n) = (x - \tilde{x}_n)f'(c_n).$$

If $f'$ is continuous on $[a, b]$, then:

$$\sup_{x \in [a,b]} |f'(x)| \le M,$$

and therefore:

$$|f(x) - f(\tilde{x}_n)| \le M|x - \tilde{x}_n| \le M\delta \le \frac{M(b-a)}{N}.$$

Thus:

$$E_N = |\tilde{J} - J| = \left| \sum_{n=0}^{N-1} \left( \int_{x_n}^{x_{n+1}} f(x)\, dx - \delta f(\tilde{x}_n) \right) \right|$$

$$\le \sum_{n=0}^{N-1} \left| \int_{x_n}^{x_{n+1}} (f(x) - f(\tilde{x}_n))\, dx \right| \le \sum_{n=0}^{N-1} \int_{x_n}^{x_{n+1}} |f(x) - f(\tilde{x}_n)|\, dx$$

$$\le \sum_{n=0}^{N-1} \int_{x_n}^{x_{n+1}} \frac{M(b-a)}{N}\, dx \le \frac{M(b-a)^2}{N}.$$

Therefore, if $f \in C^1([a, b])$, then the rectangle rule converges and is of order one.

## SOLUTION 5.2   TRAPEZOIDAL RULE

**a.** In a similar way to the rectangle rule:

```python
def trapezes(f,a,b,N):
    delta = (b-a)/N
    J = 0
    for i in range(N):
        x = a + delta*i
        J += delta*(f(x)+f(x+delta))/2
    return J
```

This gives:

```python
trapezes(f,0,1,100)
```

**b.** For the calculation of $J_n$, $f(x_n)$ and $f(x_{n+1})$ need to be calculated, but for $J_{n+1}$, it is possible to reuse the evaluation of $f(x_{n+1})$. Summing up all $J_n$, we get:

$$\tilde{J} = \sum_{n=0}^{N-1} J_n = \frac{\delta}{2} \sum_{n=0}^{N-1} (f(x_n) + f(x_{n+1})) = \delta \left( \frac{f(x_0)}{2} + \sum_{n=1}^{N-1} f(x_n) + \frac{f(x_N)}{2} \right)$$

Thus, the following function performs only the $N + 1$ evaluations of $f$ needed and not $2N$ as with the previous version:

```
def trapezes2(f,a,b,N):
    delta = (b-a)/N
    J = delta*(f(a)+f(b))/2
    for i in range(1,N):
        x = a + delta*i
        J += delta*f(x)
    return J
```

The results are the same:

```
trapezes(f,0,1,100) - trapezes2(f,0,1,100)
```

to the rounding error.

**c.** In a similar way to the rectangle rule:

```
# list of values of N
list_N = np.arange(10,1000)
# apply the function on each element of the list
data = np.vectorize(lambda N: trapezes2(f,0,1,N))(list_N)
plt.figure(figsize=(8,5))
plt.title("Second-order convergence of the trapezoidal rule")
plt.xlabel("$N$")
plt.ylabel(r"$N^2 \times E_N$")
plt.ylim(0,1)
plt.plot(list_N, np.abs(data-Jexact)*list_N**2)
```

or in logarithmic scales:

```
plt.figure(figsize=(8,5))
plt.title("Order of convergence of the trapezoidal rule")
plt.xlabel("$N$")
plt.loglog(list_N, np.abs(data-Jexact), label="$E_N$")
plt.loglog(list_N, 1/list_N**2, label="$N^{-2}$")
plt.legend()
```

Figures 5.4 and 5.5 indicate that the speed of convergence of the trapezoidal rule is of order two: $E_N \propto N^{-2}$.

**d.** The linear function that approximates $f$ on the interval $[x_n, x_{n+1}]$ is:

$$L_n(x) = \frac{x - x_n}{\delta} f(x_{n+1}) + \frac{x_{n+1} - x}{\delta} f(x_n),$$

so that:

$$J_n = \int_{x_n}^{x_{n+1}} L_n(x)\,dx = \frac{f(x_n) + f(x_{n+1})}{2}.$$

By Lagrange's approximation theorem, for $x \in [x_n, x_{n+1}]$:

$$|f(x) - L_n(x)| \leq \frac{(x - x_n)(x_{n+1} - x)}{2} M',$$

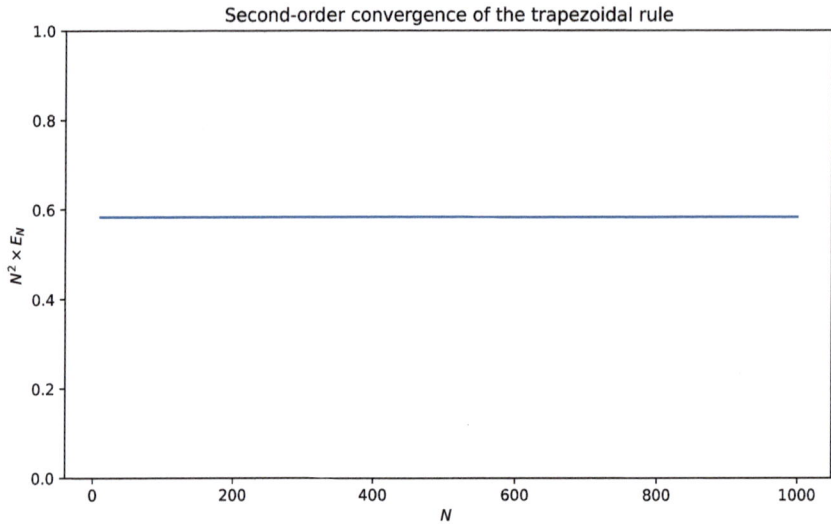

**Figure 5.4** Error $E_N = |\tilde{J} - J|$ multiplied by $N^2$ as a function of $N$ for the trapezoidal rule: this method is of order two.

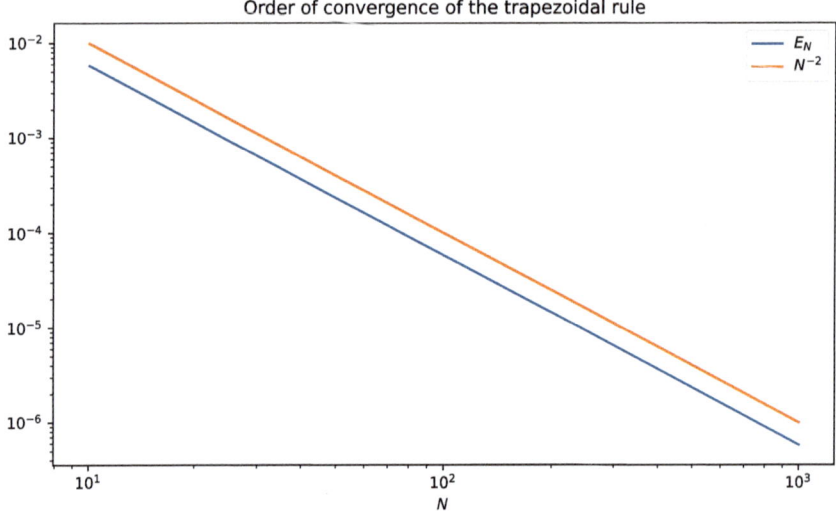

**Figure 5.5** Comparison between the error $E_N$ for the trapezoidal rule and the function $N^{-2}$: the trapezoidal rule is indeed of order two.

where

$$M' = \sup_{x \in [a,b]} |f''(x)|.$$

Thus:

$$E_N = \left| \int_a^b f(x)\,dx - \sum_{n=0}^{N-1} J_n \right| \le \sum_{n=0}^{N-1} \int_{x_n}^{x_{n+1}} |f(x) - L_n(x)|\,dx$$

$$\le \frac{M'}{2} \sum_{n=0}^{N-1} \int_{x_n}^{x_{n+1}} (x - x_n)(x_{n+1} - x) \le \frac{NM'\delta^3}{12} \le \frac{M'(b-a)^3}{12N^2},$$

Therefore, if $f \in C^2([a,b])$, the trapezoidal rule is of order two: $E_N \propto N^{-2}$.

## SOLUTION 5.3   MONTE CARLO METHOD

**a.** We use NumPy to generate $N$ random values and compute the average:

```
def montecarlo(f,a,b,N):
    # draws N values in [a,b]
    x = a + (b-a)*np.random.random(N)
    y = np.vectorize(f)(x)
    # mean
    mean = (b-a)*np.mean(y)
    return mean
```

To test:

```
montecarlo(f,0,1,10**4)
```

**b.** By adapting the previous function:

```
def montecarlo(f,a,b,N):
    # draws N values in [a,b]
    x = a + (b-a)*np.random.random(N)
    y = np.vectorize(f)(x)
    # mean
    mean = (b-a)*np.mean(y)
    # variance
    var = (b-a)**2*np.var(y)
    return (mean,var)
```

To test:

```
montecarlo(f,0,1,10**4)
```

**c.** First, we define a function that allows us to perform statistics on $k$ evaluations of montecarlo(f,a,b,N) by returning the mean of the $\tilde{J}$ values found as well as the variance:

```python
def stats(f,a,b,N,k):
    # list of nb results of Monte Carlo
    lst = np.zeros(k)
    for i in range(k):
        lst[i],_ = montecarlo(f,0,1,N)
    return np.array([np.mean(lst), np.var(lst)])
```

Using this function, it is possible to determine the speed of convergence of the mean and variance of $\tilde{J}$ over $k$ runs as a function of $N$:

```python
Nmax = 1000; k = 100
# list of values of N
N = np.arange(1,Nmax)
# arrays for the mean and variance
mean = np.zeros(len(N))
var = np.zeros(len(N))
for i in range(len(N)):
    mean[i],var[i] = stats(f,0,1,N[i],k)
# figure
plt.figure(figsize=(8,5))
plt.title(f"Convergence of the Monte Carlo method for $k =
    {k}$ evaluations")
plt.plot(3*N**(1/2)*np.abs(mean-Jexact),
    label=r"$3N^{1/2}\,|\mathrm{\mathbb{E}}(\tilde{J})-J|$")
plt.plot(N*var, label=r"$N\,\mathrm{Var}(\tilde{J})$")
plt.xlabel("$N$")
plt.ylim(0,1)
plt.legend()
```

Figure 5.6 suggests that the mean of $\tilde{J}$ converges to $J$ as $|\mathbb{E}(\tilde{J}) - J| \propto N^{-1/2}$ and that the variance converges as $\mathrm{Var}(\tilde{J}) \propto N^{-1}$.

The previous implementation is actually not very clever because for each value of $N$ between 1 and $N_{max}$, new random numbers are generated and the function is evaluated on them. Another approach is to generate $N_{max} \times k$ random values, evaluate $f$ on them, and then take into account only $N$ of them to compute the dependence in $N$. The realizations on the different values of $N$ are obviously no longer independent with this approach. Moreover, it would be better to compute the average of the errors $E_N = |\tilde{J} - J|$ over $k$ evaluations, rather than the deviation of the average of $\tilde{J}$ from the exact value $J$. The following function returns the mean $\mathbb{E}(\tilde{J})$, the variance $\mathrm{Var}(\tilde{J})$, and the mean error $\mathbb{E}(E_N)$ over $k$ evaluations:

```python
def allstats(f,a,b,Nmax,k):
    fv = np.vectorize(f)
    # define an array of size Nmax x nb with f(v) for v in
        [a,b]
    x = fv(a + (b-a)*np.random.random((Nmax, k)))
    # calculate the cumulative sum divided by 1/N for each
        column
```

```
y = 1/np.arange(1, Nmax+1)[:, None] * np.cumsum(x,
  ↳ axis=0)
# calculate the mean on each column
mean = (b-a)*np.mean(y, axis=1)
# calculate the variance on each column
var = (b-a)**2*np.var(y, axis=1)
# calculate the mean error on each column
error = np.mean(np.abs((b-a)*y-Jexact), axis=1)
return (mean,var,error)
```

This allows to make statistics on many more realizations as represented in Figure 5.7 or 5.8 and to make sure that the convergence speeds found previously are correct:

```
Nmax = 10**5; k = 100;
# values of N
N = np.arange(1,Nmax+1)
# means, variances and errors associated
mean,var,error = allstats(f,0,1,Nmax,k)
# figure
plt.figure(figsize=(8,5))
plt.title(f"Convergence of the Monte Carlo method on $k =
  ↳ {k}$ evaluations")
plt.plot(N**(1/2)*error,
  ↳ label=r"$N^{1/2}\,\mathrm{\mathbb{E}}(E_N)$")
plt.plot(var*N, label=r"$N\,\mathrm{Var}(\tilde{J})$")
plt.xlabel("$N$")
plt.ticklabel_format(style='sci', axis='x', scilimits=(0,0))
plt.ylim(0,1)
plt.legend()
```

or in logarithmic scales:

```
plt.figure(figsize=(8,5))
plt.title(f"Order of convergence of the Monte Carlo method on
  ↳ $k={k}$ evaluations")
plt.xlabel("$N$")
plt.loglog(N, error, label="$\mathrm{\mathbb{E}}(E_N)$")
plt.loglog(N, var, label=r"$\mathrm{Var}(\tilde{J})$")
plt.loglog(N, 1/N**(1/2), label="$N^{-1/2}$")
plt.loglog(N, 1/N, label="$N^{-1}$")
plt.legend()
```

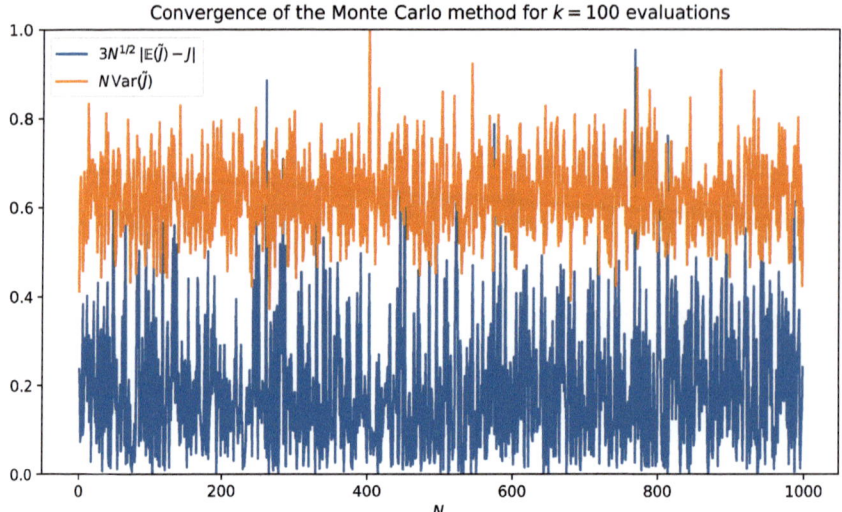

Figure 5.6 Representation of the convergence of the mean of the $\tilde{J}$ to the exact value $J$ as well as the variance of the $\tilde{J}$ over $k = 100$ runs of the Monte Carlo method.

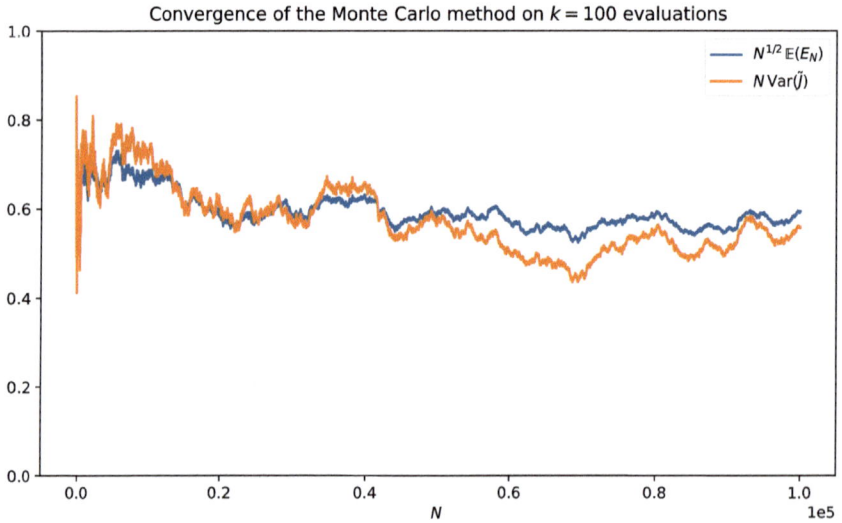

Figure 5.7 Statistics on 100 realizations as a function of $N$ of the mean of the errors $E_N = |\tilde{J} - J|$ and of the variance of $\tilde{J}$. It is clear that the convergence of the mean of the errors is in $N^{-1/2}$ and that of the variance is in $N^{-1}$.

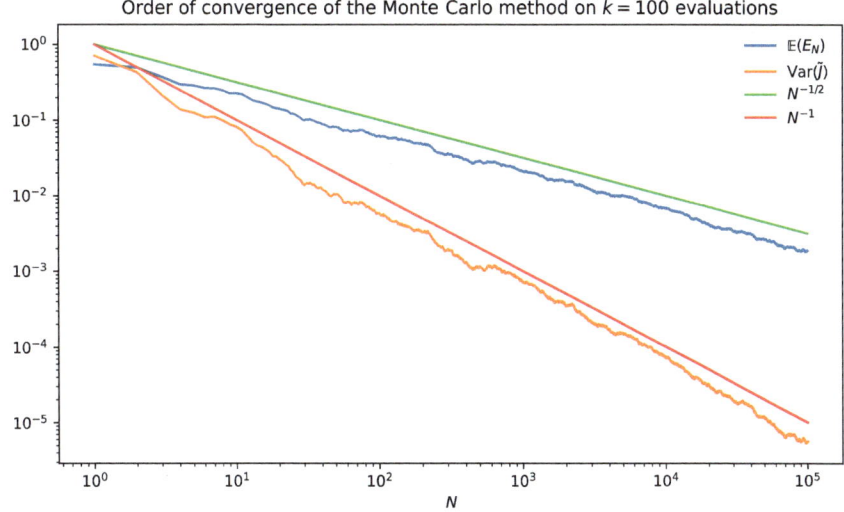

Figure 5.8    In logarithmic scales, the orders of convergence of the mean $E_N$ errors and the variance of $\tilde{J}$ are even more obvious.

**d.** According to the central limit theorem, if $Y_i$ is a sequence of independent random variables of expectation $\mu$ and variance $\sigma^2$, then the random variable:

$$S_N = \frac{1}{N} \sum_{i=0}^{N-1} Y_i,$$

has an expectation $\mu$ and a variance:

$$\text{Var}(S_N) = \frac{\sigma^2}{N}.$$

Taking $Y_i = f(X_i)$ with $X_i$ a sequence of independent random variables uniformly distributed on $[a, b]$, then the expectation of $Y_i$ is the mean of $f$ and thus the expectation of $S_N$ is given by:

$$\mathbb{E}(S_N) = \frac{1}{b-a} \int_a^b f(x)\,dx.$$

The variance of $Y_i$ is also the variance of $f$, $\sigma^2 = \text{Var}(f(X))$ and so the variance of $S_N$ is:

$$\text{Var}(S_N) = \frac{\text{Var}(f(X))}{N}.$$

Therefore, this shows that $\tilde{J}$ converges to $J$ as $N^{-1/2}$ since the variance is proportional to $N^{-1}$.

Note that to establish this result, no regularity condition on $f$ is necessary, integrability is enough.
In practice, the variance of $f$ can be estimated by the empirical variance.

**Remark:** Although not very useful in one dimension, the Monte Carlo method is extremely efficient for computing integrals in high dimensions because the speed of convergence of this method is independent of the dimension.

## SOLUTION 5.4   SIMPSON'S RULE (!)

**a.** Given the indication, noting $m_n = \frac{x_n + x_{n+1}}{2}$, then:

$$p_n(x) = \frac{(x - m_n)(x - x_{n+1})}{(x_n - m_n)(x_n - x_{n+1})} f(x_n) + \frac{(x - x_n)(x - x_{n+1})}{(m_n - x_n)(m_n - x_{n+1})} f(m_n)$$
$$+ \frac{(x - x_n)(x - m_n)}{(x_{n+1} - x_n)(x_{n+1} - m_n)} f(x_{n+1}).$$

**b.** Integrating $p_n$ by hand, we find:

$$J_n \approx \int_{x_n}^{x_{n+1}} p_n(x)\, dx$$

$$\approx \frac{x_{n+1} - x_n}{6} f(x_n) + \frac{2(x_{n+1} - x_n)}{3} f(m_n) + \frac{x_{n+1} - x_n}{6} f(x_{n+1})$$

$$\approx \frac{\delta}{6} \left( f(x_n) + 4f(m_n) + f(x_{n+1}) \right).$$

It is also possible to do it automatically with SymPy:

```python
import sympy as sp
sp.init_printing()
# define the symbols
func = sp.Function("f")
x = sp.Symbol("x")
xn = sp.Symbol("x_{n}")
xp = sp.Symbol("x_{n+1}")
m = (xn+xp)/2
# define the polynomial
pn = (x-m)*(x-xp)/(xn-m)/(xn-xp)*func(xn) +
    (x-xn)*(x-xp)/(m-xn)/(m-xp)*func(m) +
    (x-xn)*(x-m)/(xp-xn)/(xp-m)*func(xp)
# calculate and simplify the integral
integral = sp.simplify(sp.integrate(pn,(x,xn,xp)))
sp.simplify(integral.subs(xp,xn+sp.Symbol(r"\delta")))
```

**c.** By summing the approximations of $J_n$:

$$\tilde{J} = \frac{\delta}{6} \sum_{n=0}^{N-1} [f(x_n) + 4f(m_n) + f(x_{n+1})]$$

$$= \frac{\delta}{6} \left[ \sum_{n=0}^{N-1} f(x_n) + 4 \sum_{n=0}^{N-1} f(m_n) + \sum_{i=1}^{N} f(x_n) \right]$$

$$= \frac{\delta}{6} \left[ \sum_{n=0}^{N-1} f(x_n) + 4 \sum_{n=0}^{N-1} f(m_n) + \sum_{n=0}^{N-1} f(x_n) - f(a) + f(b) \right]$$

$$= \frac{\delta}{3} \left[ \frac{f(b) - f(a)}{2} + \sum_{n=0}^{N-1} \left( f(x_n) + 2f \left( \frac{x_n + x_{n+1}}{2} \right) \right) \right].$$

**d.** The implementation of the formula is straightforward:

```python
def simpson(f,a,b,N):
    delta = (b-a)/N
    J = delta/6*(f(b)-f(a))
    for i in range(0,N):
        x = a + delta*i
        J += delta/3*f(x) + 2*delta/3*f(x+delta/2)
    return J
```

To test:

```python
simpson(f,0,1,10)
```

Note that in the following example, Simpson's rule is accurate to the nearest numerical precision:

```python
simpson(lambda x: 3*x**2 + 2*x - 1 + np.sin(2*np.pi*x), 0, 1,
    10) - 1
```

because it is the sum of a polynomial of order two and an odd function with respect to the middle of the interval $[0, 1]$.

**e.** The following code allows to compare the different methods as a function of $N$ for a smooth function and a function that is only continuous but not derivable as shown in Figure 5.9:

```python
# create a new figure with a title
fig = plt.figure(figsize=(14,5))
fig.suptitle(r"Comparison of numerical integration methods")
# list of the values of N
list_N = np.arange(1,1000)
# functions and methods
functions = [{"f": f, "J": Jexact, "label": "smooth"}, \
```

```
                     {"f": lambda x: np.sqrt(1-x**2), "J": np.pi/4,
                      ↳ "label": "non-smooth"},]
methods = [(lambda *args: rectangles(*args,alpha=0),
    ↳ r"Rectangle rule with $\alpha=0$"), \
              (trapezes2, "Trapezoidal rule"), \
              (simpson,"Simpson's rule")]
for i,dic in enumerate(functions):
    # create a subfigure
    sub = fig.add_subplot(1,2,i+1)
    sub.set_title(f"for a {dic['label']} function")
    plt.xlabel("$N$")
    plt.ylabel(r"$E_N$")
    plt.ylim(1e-16,2)
    # iterate over the methods
    for method,label in methods:
        data = np.vectorize(lambda N:
          ↳ method(dic['f'],0,1,N))(list_N)
        plt.loglog(list_N, np.abs(data-dic['J']),
          ↳ label=label)
    plt.legend()
```

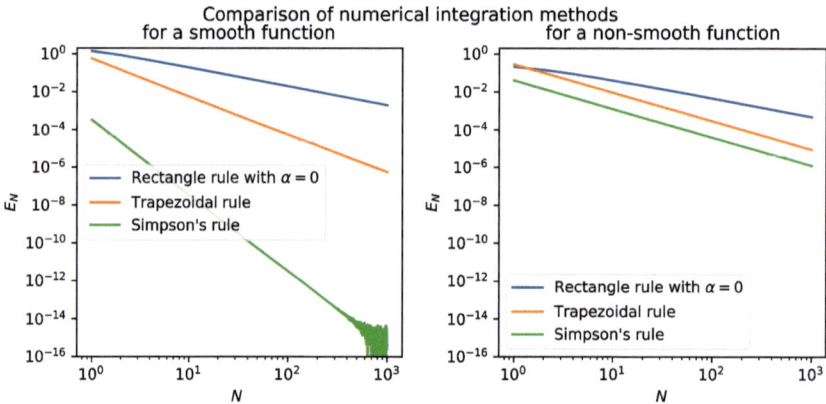

Figure 5.9   Comparison between the rectangle, trapezoidal, and Simpson's rules. For the smooth function, the orders of convergence are respectively $N^{-1}$, $N^{-2}$, and $N^{-3}$. On the one hand, for the function $f(x) = \sqrt{1 - x^2}$ which is not derivable in $x = 1$, the method of rectangles converges well as $N^{-1}$. On the other hand, the two other methods converge only as $N^{-3/2}$ visibly: the order of convergence is limited by the regularity of the function. The oscillations observed when $N$ is large on the curve of the Simpson's rule come from the limit of representation of the floating numbers which is about $10^{-16}$.

**f.** The idea is to compute the points where the function will be evaluated, *i.e.*, on the $2N + 1$ evenly spaced points $(x_0, m_0, x_1, m_1, x_{N-1}, m_{N-1}, x_N)$. It is then a matter of using slicing to select the odd points (the $x_n$) or the even ones (the $m_n$):

```python
def simpson2(f,a,b,N):
    delta = (b-a)/N
    # points where f is evaluated
    x = np.linspace(a,b,2*N+1)
    # evaludate f on these points
    y = f(x)
    # Simpson sum
    J = np.sum(y[:-2:2] + 4*y[1:-1:2] + y[2::2])
    J *= delta/6
    return J
```

For example, for N=1000, this implementation is about 50 times faster.

## SOLUTION 5.5   INTEGRATION WITH SCIPY (‼)

**a)** The example is in the documentation; it is a matter of defining the integrator and then passing the fixed parameters to the quad function:

```python
import math, scipy.integrate
def integrand(t, n, x):
    return math.exp(-x*t)/t**n
def E(n, x):
    return scipy.integrate.quad(integrand, 1, math.inf,
    ↪    args=(n, x))[0]
E(4,2)
```

**b)** The function dblquad of SciPy allows to calculate directly the requested integral:

```python
scipy.integrate.dblquad(lambda x, y: x*math.sin(x*y), 0,
↪    math.pi, 0, lambda x: x)
```

# CHAPTER 6

# Algebra

First, a method for solving a linear system by a direct algorithm is studied; then, an iterative method will be used to determine the eigenvector associated to the largest eigenvalue of a matrix. Finally, the groups generated by a set of permutations will be studied.

Concepts covered

- direct solver (LU decomposition)

- *in place* algorithm

- iterative solver (iterated power)

- groups of permutations

- orbit and stabilizer

DOI: 10.1201/9781003565451-6

# EXERCISES

## EXERCISE 6.1   LU DECOMPOSITION

The LU decomposition consists in decomposing a matrix $A$ of size $n \times n$ into the form $A = LU$, where $L$ is a lower triangular matrix with 1 on the diagonal and $U$ an upper triangular matrix. Once the decomposition $A = LU$ of a matrix is known, it is then very easy to solve the linear problem $Ax = b$ by solving first $Ly = b$ then $Ux = y$. The advantage of the LU decomposition over the Gauss algorithm, for example, is that once the LU decomposition is known, it is possible to solve the linear system quickly with different right-hand sides.

Since $l_{ik} = 0$ if $k > i$, we have:

$$a_{ij} = \sum_{k=1}^{n} l_{ik} u_{kj} = l_{ii} u_{ij} + \sum_{k=1}^{i-1} l_{ik} u_{kj},$$

and therefore as $l_{ii} = 1$:

$$u_{ij} = a_{ij} - \sum_{k=1}^{i-1} l_{ik} u_{kj}.$$

On the contrary, since $u_{kj} = 0$ if $k > j$, then:

$$a_{ij} = \sum_{k=1}^{n} l_{ik} u_{kj} = l_{ij} u_{jj} + \sum_{k=1}^{j-1} l_{ik} u_{kj},$$

and therefore if $u_{jj} \neq 0$:

$$l_{ij} = \frac{1}{u_{jj}} \left( a_{ij} - \sum_{k=1}^{j-1} l_{ik} u_{kj} \right).$$

Thus, if the first $(i-1)$ rows of $U$ and the first $(i-1)$ columns of $L$ are known, it is possible to determine the $i$-th row of $U$ by:

$$u_{ij} = a_{ij} - \sum_{k=1}^{i-1} l_{ik} u_{kj}, \quad j \geq i,$$

then, the $i$-th column of $L$ by:

$$l_{ji} = \frac{1}{u_{ii}} \left( a_{ji} - \sum_{k=1}^{i-1} l_{jk} u_{ki} \right), \quad j > i.$$

This algorithm for the LU decomposition of a matrix $A$ requires that $u_{ii}$ is never zero. This is indeed the case when the matrix $A$ and all its principal submatrices are invertible.

**a.** Write a function LU(A) that returns the result of the LU decomposition of a matrix.

**b.** Given the LU decomposition of a matrix $A$, write a function solve(L,U,b) that solves the linear system $Ax = b$.

**c.** Write a function LU_inplace(A) that does not create new arrays for $L$ and $U$ but modifies $A$ so that its lower triangular part (without the diagonal) matches $L$ and its upper triangular part (with the diagonal) matches $U$. Also make a version solve_inplace that takes as input the output of LU_inplace and returns the solution $x$ without using any new arrays.

**d.** Using the LU decomposition of the matrix $A$, write a function det(A) that returns its determinant.

## EXERCISE 6.2   POWER ITERATION METHOD

The goal of this exercise is to determine the eigenvector of a matrix associated to the largest eigenvalue (in modulus), assuming that this one is unique. Given a real matrix $A$ of size $n \times n$ and a vector $x_0 \in \mathbb{R}^n$, the sequence of vectors $(x_k)_{k \in \mathbb{N}}$ is defined by:

$$x_{k+1} = \frac{Ax_k}{\|Ax_k\|}.$$

Assuming, for example, that the matrix $A$ is diagonalizable, it is then possible to show that the sequence $(x_k)_{k \in \mathbb{N}}$ converges up to a sign to the eigenvector associated to the largest eigenvalue of $A$ in absolute value. The convergence takes place almost surely for all choices of $x_0$.

**a.** Define a function power(A, x0, k) that returns $x_k$. With this function, determine the largest eigenvector of the matrix:

$$A = \begin{pmatrix} 0.5 & 0.5 \\ 0.2 & 0.8 \end{pmatrix}.$$

*Answer: The largest eigenvector is $\pm(0.70710678, 0.70710678)$.*

**b.** Determine the eigenvalue associated with the previous eigenvector.
*Hint: If $v$ is a normalized eigenvector of $A$, then the associated eigenvalue is given by $\lambda = Av \cdot v$.*

**c.** Write a function to automatically compute the largest eigenvalue (in modulus) and the associated eigenvector of a square matrix with a given precision. We will choose the initial vector $x_0$ randomly.

**d.** Assuming the matrix $A$ is diagonalizable with a single eigenvalue of largest modulus, show that the sequence $x_k$ converges up to one sign to the eigenvector associated with this largest eigenvalue for almost all choices of $x_0$.
*Hint: Decompose $x_0$ in the eigenvector basis of A. For simplicity, we can assume that the eigenvalue of largest modulus is positive.*

**e.** Look at the NumPy documentation to find the functions to compute the eigenvectors and eigenvalues of a matrix.

**f.** Compare the speed of the previous code and the NumPy functions for matrices of size $n \times n$ with $n = 10, 100, 1\,000$.
*Hint: Taking for example matrices whose coefficients are randomly generated in the interval $(0, 1)$, the Perron-Frobenius theorem ensures the existence of a single eigenvalue of maximum modulus that is positive.*

## EXERCISE 6.3   EXPONENTIAL OF MATRICES

The goal of this exercise is to develop an algorithm to calculate the exponential of a real square matrix. If $A$ is a real square matrix, its exponential is defined by the series:

$$\exp(A) = \sum_{k=0}^{\infty} \frac{A^k}{k!},$$

by analogy with the exponential on real numbers. Here, $A^k$ represents the matrix product of $A$ with itself $k$ times.

**a.** Define the NumPy arrays corresponding to the matrices $A_1$ and $A_2$ defined by:

$$A_1 = \begin{pmatrix} 1 & 0.8 & 0.6 \\ 0.8 & 0.2 & 0.8 \\ 0 & 1.2 & 0.9 \end{pmatrix}, \qquad A_2 = \begin{pmatrix} 2 & 3 & 2 \\ 1 & 2 & 3 \\ 4 & 3 & 5.2 \end{pmatrix}.$$

**b.** Define a function `matrix_power(A,n=20)` returning an approximation of $\exp(A)$ obtained by keeping only the first $n + 1$ terms of the series, *i.e.*, the terms from $k = 0$ to $k = n$.

**c.** Test on the matrices $A_1$ and $A_2$ and compare with the results of the function `expm` of the module `linalg` of SciPy.

**d.** Without using the function `norm` of NumPy or SciPy, define a function computing the Frobenius norm $\|A\|_F$ of a matrix $A$ of size $m \times m$ defined by:

$$\|A\|_F^2 = \mathrm{tr}(AA^t) = \sum_{i=1}^{m} \sum_{j=1}^{m} |a_{ij}|^2.$$

Compute the Frobenius norms of the matrices $A_1$ and $A_2$.

**e.** For matrices $A_1$ and $A_2$, plot the error in the Frobenius norm between the result of `matrix_power` and the result of `expm` as a function of the number of terms $n$ kept. Put a logarithmic scale on the y-axis.
From a theoretical point of view, it is possible to show that the error is bounded by:

$$\left\| \exp(A) - \sum_{k=0}^{n} \frac{A^k}{k!} \right\|_F \leq \frac{e^{\|A\|_F}}{(n+1)!} \|A\|_F^{n+1}.$$

**f.** Plot this bound as a function of $n$ for different values of the $\|A\|_F$ ranging from 2 to 20, with also a logarithmic scale on the y-axis. Roughly deduce the number of terms to keep so that the bound is lower than the machine precision of $10^{-15}$ if $\|A\|_F = 20$. Compare the theoretical bound with what was observed in the previous question.

A basic idea to improve the convergence of the series when the norm of the matrix is large is to perform a rescaling using the relation:

$$\exp A = \left(\exp(A/p)\right)^p,$$

for $p \geq 1$, a well-chosen integer such that $\|A/p\|_F$ is small, for example, less than one.

**g.** Using the previous property, write a function `matrix_power_opt(A,n=20)` based on this property.

**h.** Redo the same graph as at point **e** but with this new function and comment.

## EXERCISE 6.4   GROUPS OF PERMUTATIONS

The goal is to study the groups of permutations by developing algorithms to characterize them. A group of permutations $G \subset S_n$ is defined as being generated by a number of permutations: $G = \langle g_1, g_2, \dots, g_k \rangle$, with $g_i \in S_n$ a permutation of the set $\{1, 2, \dots, n\}$. A permutation:

$$g = \begin{pmatrix} 1 & 2 & 3 & 4 \\ 4 & 1 & 2 & 3 \end{pmatrix},$$

can be represented in Python by the tuple `g = (0, 4, 1, 2, 3)`. The zero is added in order to conform with Python's zero-based indexing and thus simplify the implementations a bit. This simply means that vertex 0 goes on vertex 0. Note that this exercise lends itself particularly well to an object-oriented implementation, and in this case the questions can be adapted accordingly.

**a.** Write a function `product(g1,g2)` that computes the product of two permutations.

**b.** Write a function `inverse(g)` that computes the inverse of a permutation.

**c.** Write a function that returns the decomposition of a permutation into cycles represented by a list of tuples.

**d.** Write a function that returns the permutation corresponding to a list of cycles, *i.e.*, that does the inverse of the previous function.

**e.** ! In Python, a group $G = \langle g_1, g_2, \dots, g_k \rangle$ generated by a family of permutations can be represented by a list of permutations $G = [g1,g2, \dots ,gk]$. Write a function `orbit(G,x)` that returns the orbit of a point $x \in \{1, 2, \dots, n\}$:

$$O_x = Gx = \{gx, g \in G\}.$$

*Hint: Do not compute the set of elements of the group, it makes a list much too long. Construct the list $(X^0, X^1, X^N)$ of disjoint sets defined recursively by $X^0 = \{x\}$ and $X^n$ as the set of new elements of the form $g_i y$ with $1 \le i \le k$ and $y \in X^{n-1}$:*

$$X^n = \left( \bigcup_{i=1}^{k} g_i X^{n-1} \right) \setminus \left( \bigcup_{i=1}^{n-1} X^i \right).$$

**f.** ‼ Define a function `stabilizer(G,x)` that returns the stabilizer of a point $x \in \{1, 2, \dots, n\}$:

$$G_x = \{g \in G : gx = x\},$$

in the form of a set of generators.
*Hint: Use Schreier's lemma.*

# SOLUTIONS

## SOLUTION 6.1   LU DECOMPOSITION

**a.** Applying the previous formulas and taking care that the 1 on the diagonal of $L$ must be forced:

```python
import numpy as np
def LU(A):
    # dimensions
    n = A.shape[0]
    # initialization of the matrices L and U
    L = np.identity(n)
    U = np.zeros((n,n))
    # loop on i i
    for i in range(n):
        # calculate the i-th row of U
        for j in range(i,n):
            U[i,j] = A[i,j] - np.dot(L[i,:],U[:,j])
        # stop if U_ii is too small
        if abs(U[i,i]) < 1e-16:
            raise Exception("Zero pivot")
        # calculate the i-th column of L
        for j in range(i+1,n):
            L[j,i] = (A[j,i] - np.dot(L[j,:],U[:,i]))/U[i,i]
    return (L,U)
```

We check that the product does what we want:

```python
A = np.random.random((10,10))
L,U = LU(A)
np.linalg.norm(A - L @ U)
```

**b.** We first solve $Ly = b$ by:

$$y_i = b_i - \sum_{j=1}^{i-1} l_{ij} y_j,$$

then $Ux = y$ by:

$$x_i = \frac{1}{u_{ii}} \left( y_i - \sum_{j=i+1}^{n} u_{ij} x_j \right).$$

The implementation is as follows:

```
def solve(L,U,b):
    n = L.shape[0]
    # solve Ly = b
    y = np.zeros(n)
    for i in range(n):
        y[i] = b[i] - np.dot(L[i,:],y)
    # solve Ux = y
    x = np.zeros(n)
    for i in reversed(range(n)):
        x[i] = (y[i] - np.dot(U[i,:],x))/U[i,i]
    return x
```

and it returns the right solution:

```
A = np.random.random((4,4))
b = np.random.random(4)
L,U = LU(A)
A @ solve(L,U,b) - b
```

**c.** The idea is exactly the same, but one just has to be careful to restrict the indices:

```
def LU_inplace(A):
    # dimensions
    n = A.shape[0]
    for i in range(n):
        # calculate the i-th row ofU
        for j in range(i,n):
            A[i,j] = A[i,j] - np.dot(A[i,0:i],A[0:i,j])
        # stop if A_ii is too small
        if abs(A[i,i]) < 1e-16:
            raise Exception("Zero pivot")
        # calculate the i-th column of L
        for j in range(i+1,n):
            A[j,i] = (A[j,i] -
            ↪ np.dot(A[j,0:i],A[0:i,i]))/A[i,i]
    return A
```

Note that since the matrix A is modified *in place*, it is not necessary for the previous function to return it. You have to be very careful that the matrix A is not the original matrix anymore. We check that the upper and lower triangular parts are identical to those calculated previously:

```
A = np.random.random((4,4))
L,U = LU(A)
A = LU_inplace(A)
A-L, A-U
```

To avoid creating new arrays for $x$ and $y$, it is possible to do everything in $b$, given a matrix $A$ returned by solve_inplace:

```
def solve_inplace(A,b):
    n = A.shape[0]
    # solve Ly = b in place
    for i in range(n):
        b[i] = b[i] - np.dot(A[i,:i],b[:i])
    # solve Ux = y in place
    for i in reversed(range(n)):
        b[i] = (b[i] - np.dot(A[i,i+1:],b[i+1:]))/A[i,i]
    return b
```

To test:

```
A = np.random.random((4,4))
b = np.random.random(4)
# to check the correctness of the results, one keep the
↳ values of A and b (which will be overwritten)
Akeep = A.copy()
bkeep = b.copy()
# solve in place and overwrite A and b
A = LU_inplace(A)
Akeep @ solve_inplace(A,b) - bkeep
```

**d.** Since $L$ is lower triangular with 1 on the diagonal:

$$\det(A) = \det(L)\det(U) = \det(U) = \prod_{i=1}^{n} u_{ii} ,$$

and therefore using either the traditional method or the *in place* method:

```
def det(A):
    L,U = LU(A)
    return U.diagonal().prod()
def det_inplace(A):
    LU_inplace(A)
    return A.diagonal().prod()
```

## SOLUTION 6.2   POWER ITERATION METHOD

**a.** It is sufficient to act k times with the matrix A and to normalize at each step by the norm:

```
def power(A, x0, k):
    # define a new vector initialized with x0
    xk = x0.copy()
    for _ in range(k):
        # calculate the matrix product of A with xk
        xk = np.dot(A, xk)
        # calculate the norm
```

```
        norm = np.sqrt(np.dot(xk,xk))
        # normalize the vector xk
        xk /= norm
    return xk
```

In order not to find a particular case for which the method would not converge, the simplest way is to draw the vector $x_0$ randomly:

```
x0 = np.random.random(2)
A = np.array([[0.5, 0.5], [0.2, 0.8]])
v1 = power(A, x0, 100)
```

which gives approximately the vector (0.70710678, 0.70710678).

**b.** The simplest way to calculate the eigenvalue associated with a given normalized eigenvector is to take the scalar product:

```
lambda1 = np.dot(A @ v1,v1)
A @ v1 - lambda1*v1
```

**c.** The idea is to put everything together and stop the iterations when the required precision is reached:

```
def maxeig(A, precision=1e-8, maxiter=1000, verbose=False):
    # initialization with a random vector
    x = np.random.random(A.shape[1])
    # fix a maximal number of iterations in case the method
    ↵  diverges
    for i in range(maxiter):
        # calculate the matrix product of A with x
        Ax = np.dot(A, x)
        # determine the associated eigenvalue
        val = np.dot(Ax, x)
        # determine the error
        error = np.linalg.norm(Ax - val*x)
        # exit the loop if enough iterations
        if error < precision:
            if verbose: print(f"Converged in {i} iterations")
            break
        # new iteration
        norm = np.linalg.norm(Ax)
        x = Ax/norm
    # is not converged
    if verbose and i == maxiter-1:
        print(f"Not converged within {maxiter} iterations,
        ↵  the error is {error}.")
    return (val,x)
```

**d.** The first step is to notice that:

$$x_k = \frac{A^2 x_{k-2}}{\|A^2 x_{k-2}\|} = \cdots = \frac{A^k x_0}{\|A^k x_0\|}.$$

Since $A$ is diagonalizable, let $(v_1, v_2, \ldots, v_n)$ be a basis of eigenvectors of $A$ associated to eigenvalues $\lambda_1, \lambda_2, \ldots, \lambda_n$. Without loss of generality, we assume that $\lambda_1$ is the eigenvalue of largest modulus, i.e., $|\lambda_1| > \max(|\lambda_2|, |\lambda_3|, \ldots, |\lambda_n|)$. Note that this implies that $\lambda_1$ is real. The vector $x_0$ decomposes into the basis:

$$x_0 = \sum_{i=1}^{n} c_i v_i,$$

thus assuming that $c_1 \neq 0$:

$$A^k x_0 = \sum_{i=1}^{n} c_i \lambda_i^k v_i = c_1 \lambda_1^k \left( v_1 + \sum_{i=2}^{n} \frac{c_i}{c_1} \left( \frac{\lambda_i}{\lambda_1} \right)^k v_i \right).$$

Since $|\lambda_1| > |\lambda_i|$ for $i \geq 2$, then:

$$\lim_{k \to \infty} \left( \frac{A^k x_0}{\lambda_1^k} \right) = c_1 \lim_{k \to \infty} \left( v_1 + \sum_{i=2}^{n} \frac{c_i}{c_1} \left( \frac{\lambda_i}{\lambda_1} \right)^k v_i \right) = c_1 v_1,$$

since $\left| \frac{\lambda_i}{\lambda_1} \right| < 1$. Therefore,

$$\lim_{k \to \infty} \operatorname{sign}(\lambda_1)^k x_k = \lim_{k \to \infty} \left( \frac{|\lambda_1|}{\lambda_1} \right)^k \frac{A^k x_0}{\|A^k x_0\|} = \operatorname{sign} c_1 \frac{v_1}{\|v_1\|}.$$

By choosing $x_0$ randomly, then $c_1 \neq 0$ almost surely and therefore the sequence $(x_k)_{k \in \mathbb{N}}$ converges up to a sign to a normalized eigenvector associated with the largest modulus eigenvalue.

**e.** The function eig returns the tuple formed by the eigenvalues and eigenvectors:

```
A = np.array([[0.5, 0.5], [0.2, 0.8]])
eigenvalues, eigenvectors = np.linalg.eig(A)
eigenvectors[:,1] # retourne the second eigenvector
```

**f.** To determine the speed of the two methods:

```
%%timeit
n=10
A = np.random.random((n,n))
np.linalg.eig(A)
```

and:

```
%%timeit
n=10
A = np.random.random((n,n))
maxeig(A)
```

The results are as follows:

| $n$ | np.linalg.eig | maxeig |
|---|---|---|
| 10 | 225 μs ± 9.76 μs | 811 μs ± 74.4 μs |
| 100 | 22.1 ms ± 2.47 ms | 750 μs ± 100 μs |
| 1 000 | 2.01 s ± 254 ms | 27.5 ms ± 2.38 ms |

The NumPy algorithm is faster for small matrices, while it is the opposite for large matrices. The reason is that our method computes only one eigenvector, while NumPy computes them all. But for some applications, it is not useful to determine all the eigenvalues and all the eigenvectors.

## SOLUTION 6.3   EXPONENTIAL OF MATRICES

**a.** Just copy the values:

```
A1 = np.array([[1,0.8,0.6],[0.8,0.2,0.8],[0,1.2,0.9]])
A2 = np.array([[2,3,2],[1,2,3],[4,3,5.2]])
```

**b.** The idea is to define a matrix S that will hold the sum and a matrix Ak that contains $\frac{A^k}{k!}$; thus, the factorial is included in the calculation:

```
def matrix_exp(A, n=20):
    # for the cumulative sum
    S = np.identity(A.shape[0])
    # for the cumulative sum of A^k/k!
    Ak = np.identity(A.shape[0])
    # loop to perform the sum
    for k in range(1,n+1):
        Ak = np.dot(Ak,A)/k
        S += Ak
    return S
```

**c.** First of all one has to import the function expm from scipy.linalg:

```
from scipy.linalg import expm
```

This allows to compare the function matrix_exp with the function provided by SciPy for the matrix $A_1$:

```
matrix_exp(A1)-expm(A1)
```

and for the matrix $A_2$:

```
matrix_exp(A2)-expm(A2)
```

For the second matrix, the result is less precise.

**d.** It is a matter of copying the definition, without forgetting the square root:

```
def norm(A):
    return np.sqrt(np.trace(A@A.T))
```

The Frobenius norms of the two matrices are:

```
(norm(A1), norm(A2))
```

**e.** It is a matter of varying the n parameter to obtain Figure 6.1:

```
import matplotlib.pyplot as plt
plt.figure(figsize=(8,5))
plt.title("Error between matrix_exp and expm as function of
 ↪   the truncation order")
plt.xlabel("Truncation order $n$")
plt.ylabel("Empirical error in Frobenius norm")
for i,A in enumerate([A1,A2]):
    error = [norm(matrix_exp(A,n=n)-expm(A)) for n in
     ↪   range(1,51)]
    plt.semilogy(range(1,51), error, label=f"$A_{i+1}$")
plt.legend()
```

**f.** The aim is to plot the indicated bound as a function of $n$ for different values of the Frobenius norm of $A$:

```
plt.figure(figsize=(8,5))
plt.title("Theoretical bound as a function of the truncation
 ↪   order")
plt.xlabel("Truncation order $n$")
plt.ylabel("Theoretical error in Frobenius norm")
for a in np.linspace(2,20,10):
    val = [np.exp(a)*a**(n+1)/np.math.factorial(n+1) for n in
     ↪   range(101)]
    plt.semilogy(range(101), val, label=f"$\Vert
     ↪   A\Vert_F={a}$")
plt.ylim([1e-15,1])
plt.legend()
```

The conclusion of Figure 6.2 is that we have to keep more and more terms in the series when the norm of the matrix $A$ increases. When $\|A\|_F = 20$, it takes about a hundred terms. The theoretical bound is comparable to what has been observed previously, we need about 20 terms for $A_1$ whose norm is about 2 and about 50 for $A_2$ whose norm is close to 9.

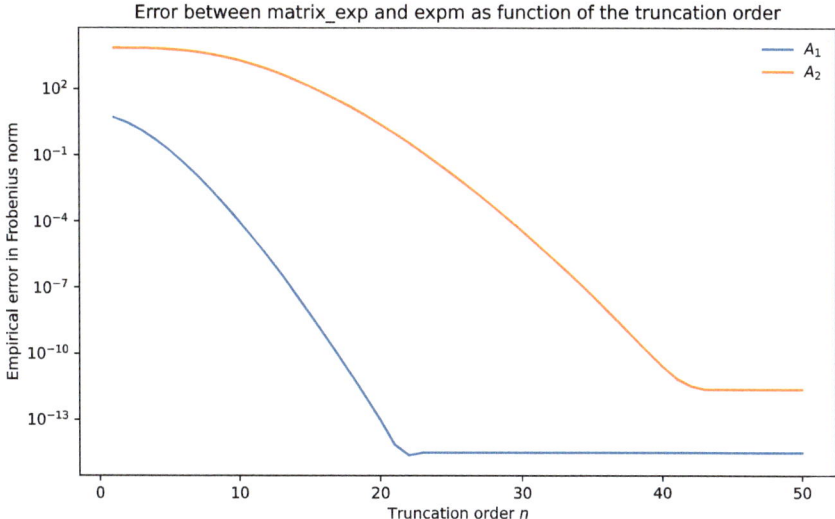

Figure 6.1    Plot in logarithmic scale of the error between our implementation of matrix exponential and the one of SciPy.

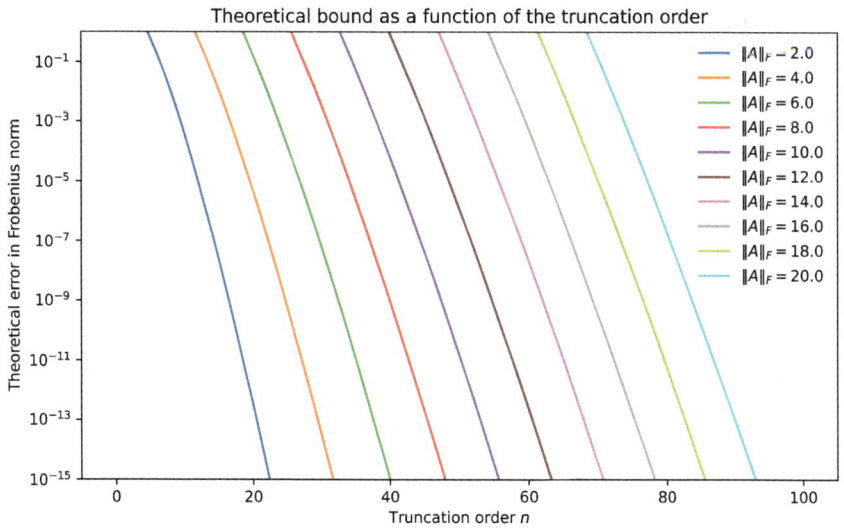

Figure 6.2    Representation of the theoretical bound on the error for matrices of various norms.

**g.** The first step is to compute $p$, then the exponential of $A/p$ as before, and finally the $p$-th power:

```python
def matrix_exp_opt(A, n=20):
    # rescaling parameter
    p = max(1,int(norm(A)))
    # rescaled matrix
    Ap = A/p
    # for the cumulative sum
    S = np.identity(A.shape[0])
    # for the cumulative sum of A^k/k!
    Ak = np.identity(A.shape[0])
    # loop to perform the sum
    for k in range(1,n+1):
        Ak = np.dot(Ak,Ap)/k
        S += Ak
    # loop to scale back
    out = np.identity(A.shape[0])
    for _ in range(p):
        out = np.dot(out,S)
    return out
```

**h)** The same code as before:

```python
plt.figure(figsize=(8,5))
plt.title("Error between matrix_exp and expm as function of
    the truncation order")
plt.xlabel("Truncation order $n$")
plt.ylabel("Empirical error in Frobenius norm")
for i,A in enumerate([A1,A2]):
    error = [norm(matrix_exp_opt(A,n=n)-expm(A)) for n in
        range(1,51)]
    plt.semilogy(range(1,51), error, label=f"$A_{i+1}$")
plt.legend()
```

allows to observe in Figure 6.3 that now the convergence takes place with only 20 terms independently of the norm of the matrix.

Note that in order to have a powerful algorithm, the second step of the $p$-th power calculation can be optimized by grouping the terms. This is particularly easy to implement if $p = 2^s$ with $s$ integer.

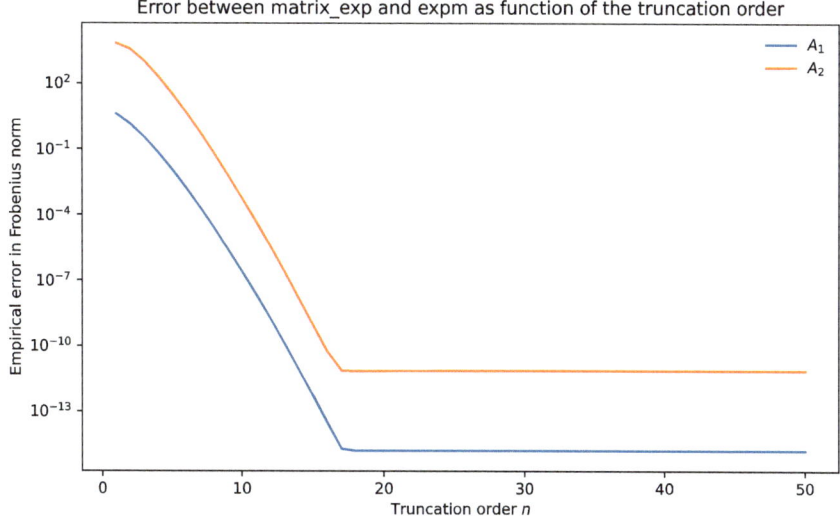

Figure 6.3   Difference between our optimized algorithm and SciPy. Our algorithm now seems to work with only 20 terms regardless of the norm of the matrix.

## SOLUTION 6.4   GROUPS OF PERMUTATIONS

**a.** This takes the composition of two permutations:

```
def product(g1,g2):
    return tuple(g1[g2i] for g2i in g2)
```

and to test:

```
g1 = (0, 5, 4, 3, 2, 1)
g2 = (0, 2, 4, 1, 3, 5)
product(g1,g2)
product(g2,g1)
```

**b.** One way to do this is to take the indexes:

```
def inverse(g):
    n = len(g)
    return tuple(g.index(i) for i in range(n))
```

which verifies the definition of the inverse:

```
product(inverse(g1),g1)
product(g1,inverse(g1))
```

**c.** The idea is to create a list with the elements that do not yet belong to a cycle:

```python
def to_cycle(g):
    n = len(g)
    # list of vertices not belonging to a cycle yet
    lst = list(range(n))
    # list of cycles
    out = []
    # classify
    while lst:
        # start a new cycle
        i = lst.pop()
        cycle = (i,)
        new = g[i]
        # end the cycle
        while new != i:
            # add to cycle
            cycle = cycle + (new,)
            # remove new from the list
            if new in lst: lst.remove(new)
            # for the next iteration
            new = g[new]
        # add the cycle to the list of cycles
        out.append(cycle)
    return out
```

To test:

```python
to_cycle(g1)
to_cycle(g2)
```

**d.** The permutation is first built as a list in order to be mutable:

```python
def to_perm(cycles):
    # dimension
    n = 0
    for c in cycles:
        n = max(n,max(c)+1)
    # permutation as list otherwise not mutable
    out = list(range(n))
    for c in cycles:
        for i in range(len(c)):
            out[c[i-1]] = c[i]
    # convert to tuple
    return tuple(out)
```

To test:

```python
to_perm(to_cycle(g1)) == g1
to_perm(to_cycle(g2)) == g2
```

**e.** The idea is to generate at each iteration the elements $g_i y$ for any $1 \leq i \leq k$ and $y$ in the list of elements generated in the previous step:

```
def orbit(G,x):
    # to store the final orbit
    out = {x}
    # to store the new elements obtained by the action of the
    ↳ generators
    Xn = {x}
    # loop while new elements are found
    while Xn:
        # set of new elements found at this iteration
        new = set()
        # loop on the generators and elements found at the
        ↳ previous iteration
        for g in G:
            for i in Xn:
                gi = g[i]
                # if not already in the orbit
                if gi not in out:
                    new.add(gi)
        # new found elements for the next iteration
        Xn = new
        # add to the final orbit
        out = out.union(Xn)
    return out
```

To test:

```
g1 = to_perm([(1, 28, 5, 3, 13, 25, 27, 8, 4, 17, 11, 29, 7,
    ↳ 2, 21, 23, 10, 6), (9, 16, 22, 30, 18, 15, 20, 19, 14)])
g2 = to_perm([(1, 4), (2, 5, 3),(30,)])
G = {g1,g2}
orbit(G,1)
```

**f.** In the previous algorithm, the elements $g_i y$ with $1 \leq i \leq k$ and $y \in X^{n-1}$ that already belong to one of the previous sets $X^i$ are clearly related to the existence of an element in the stabilizer. More precisely if:

$$z \in \left( \bigcup_{i=1}^{k} g_i X^{n-1} \right) \cap \left( \bigcup_{i=1}^{n-1} X^i \right),$$

then there exist $I = (i_1, i_2, \dots, i_n)$ and $J = (j_1, j_2, \dots, j_l)$ for some $l \leq n - 1$ such that:

$$z = g_I x, \qquad\qquad z = g_J x,$$

where

$$g_I = \prod_{i \in I} g_i, \qquad\qquad g_J = \prod_{i \in J} g_i,$$

therefore:

$$(g_J)^{-1}g_I x = x,$$

and thus $(g_J)^{-1}g_I \in G_x$. Schreier's lemma shows that $G_x$ is generated by the set of elements obtained by the previous procedure. To implement this, we need to keep in memory for each element of $X^n$ a permutation allowing to reach this element by acting on $x$. This is done by using a dictionary:

```python
def stabilizer(G,x):
    # dimension of the underlying group Sym(n)
    n=0
    for g in G:
        n = max(n,len(g))
    # identity
    identity = tuple(range(n))
    # dictionary of the form gx:g
    orb = {x:identity}
    # stabilizer
    stab = set()
    # new elements obtained by the action of the generators
    Xn = {x:identity}
    # loop while new elements are found
    while Xn:
        # new elements to be found in this iteration
        new = dict()
        # loop on the generators and elements found at the
        ↵  previous iteration
        for y in Xn:
            for i,g in enumerate(G):
                gy = g[y]
                # if already in the orbit
                if gy in orb.keys():
                    # new stabilizer
                    newstab = product(inverse(orb[gy]),
                    ↵  product(g,orb[y]))
                    stab.add(newstab)
                else:
                    # add the element not already in the
                    ↵  orbit
                    new[gy] = product(g,orb[y])
        # new found elements for the next iteration
        Xn = new
        # add to the final orbit
        orb.update(Xn)
    return stab
```

To test:

```python
for g in stabilizer(G,1):
    assert g[1] == 1
```

# CHAPTER 7

# Graph Theory

A graph is a pair $G = (X, E)$ consisting of a finite non-empty set $X$, and a set $E$ of pairs of elements of $X$. The elements of $X$ are the vertices of the graph $G$, those of $E$ are the edges of the graph $G$. A graph is oriented if the edges have a direction, *i.e.*, if the pairs of elements of $E$ are ordered lists such that $(i, j) \in E$ is not equivalent to $(j, i) \in E$. Here only undirected graphs, *i.e.*, whose pairs of elements of $X$ are unordered sets $((i, j) \in E)$, are considered.

For example, the complete graph with $n$ vertices $K_n$ is defined as the graph of vertices $X = \{1, 2, \ldots, n\}$ and of edges being the two-element of the power set of $X$. In particular, $K_4 = (X, E)$, where $X = \{1, 2, 3, 4\}$ and $E = \{\{1, 2\}, \{1, 3\}, \{1, 4\}, \{2, 3\}, \{2, 4\}, \{3, 4\}\}$.

Concepts covered

- undirected graphs

- graphs as dictionaries

- use of frozensets

- adjacency matrix

- search for paths and triangles

- recursive function

DOI: 10.1201/9781003565451-7

# EXERCISES

---

## EXERCISE 7.1   GRAPHS AS DICTIONARIES

One way to represent a graph $G$ is with a dictionary whose keys are the vertices and the value associated to each key $x \in X$ is a set containing the neighbors of $x$.

**a.** Construct the following graphs in dictionary form:

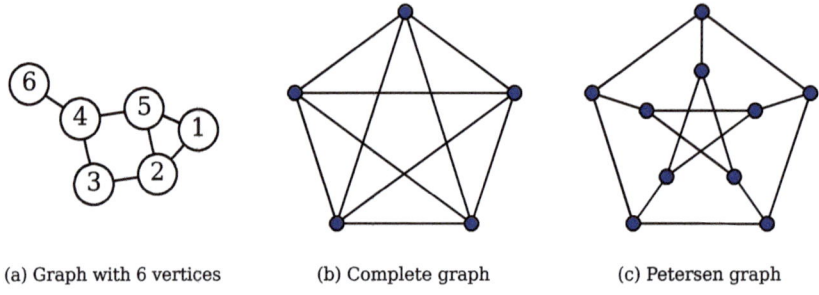

(a) Graph with 6 vertices       (b) Complete graph       (c) Petersen graph

**b.** Write a function `complete(n)` that constructs the complete graph $K_n$ as a dictionary.

**c.** A graph given as a dictionary contains the information several times. Write a function `correct(graph)` to add the missing elements of an improperly defined graph so that for any vertex `x`, if `y` belongs to `graph[x]`, then `y` is also a key and `x` belongs to `graph[y]`. Test this function, in particular, on the improperly defined graph `{1:{3,4,2},3:{2}}`.

**d.** Write a function that returns the set (type `set`) of all edges of a graph represented by a dictionary.
*Hint: Sets are mutable and therefore not hashable.*

**e.** ! Write a function to determine whether two vertices are connected by a path or not and return the path if yes.
*Hint: Use a recursive function.*

**f.** ! Write a function that returns all paths between two vertices (without cycles).

## EXERCISE 7.2   TRIANGLES IN A GRAPH

A triangle in a graph is a set of three vertices connected by three edges. Finding and analyzing triangles in a graph is important for understanding its structure.

**a.** Determine mathematically the number of subsets of cardinal three that a set of vertices $X$ has. Is it reasonable to iterate over these elements?

**b.** Write a function that returns the set of all triangles in a graph.

To each graph $G = (X, E)$ corresponds a unique symmetric matrix $A$ of size $n \times n$ with $n = |X|$ defined by:

$$A_{ij} = \begin{cases} 1, & \text{if } \{i, j\} \in E, \\ 0, & \text{if } \{i, j\} \notin E. \end{cases}$$

This matrix is called the adjacency matrix of graph $G$.

**c.** Define a function that returns the adjacency matrix of a graph.
*Hint: Be careful that the vertices are not necessarily indexed by integers between 0 and n in the dictionary.*

**d.** Define a function having as argument an adjacency matrix and returning the corresponding graph as a dictionary.

**e.** Using the adjacency matrix $A$ and the matrix $B = A^2$, write a function returning the set of triangles of a graph.

**f.** Using the adjacency matrix $A$, write a function computing the number of triangles.
*Hint: Interpret the entries of the matrix $A^n$.*

## EXERCISE 7.3   MODULE NETWORKX (!!)

Many graph theory algorithms are implemented in the NetworkX module, see the documentation at the address: https://networkx.org/documentation/.

**a.** Follow the NetworkX tutorial available at the address: https://networkx.org/documentation/stable/tutorial.html.

**b.** Analyze one of the downloadable graphs at: https://github.com/gephi/gephi/wiki/Datasets or https://snap.stanford.edu/data/.

# SOLUTIONS

## SOLUTION 7.1   GRAPHS AS DICTIONARIES

**a.** It is possible to define these graphs by hand or to use their structure to build them:

```
Ga = {1: {2,5}, 2:{1,3,5}, 3:{2,4}, 4:{3,5,6}, 5:{1,2,4},
    ↳ 6:{4}}
Gb = {i: {(a+i) % 5 for a in {1,2,3,4}} for i in range(5)}
Gc = {}
# external vertices
Gc1 = {i: {(a+i) % 5 for a in {1,4}}.union({i+5}) for i in
    ↳ range(5)}
Gc.update(Gc1)
# internal vertices
Gc2 = {i: {5 + (a+i) % 5 for a in {2,3}}.union({i-5}) for i
    ↳ in range(5,10)}
Gc.update(Gc2)
```

**b.** We build the set of all vertices, then we remove the current vertex:

```
def complet(n):
    all = set(range(0,n))
    return {i: all-{i} for i in range(n)}
```

**c.** The graph given as input must be copied to be able to modify it inside a loop:

```
def correct(graph):
    # output graph
    out = graph.copy()
    # loops on all edges
    for s in graph:
        for a in graph[s]:
            # if the corresponding key does not exist
            if a not in out:
                # add the key and corresponding value
                out[a] = {s}
            # if the corresponding key exists
            else:
                # add the value
                out[a].add(s)
    return out
```

To test:

```
correct({1:{3,4,2},3:{2}})
```

**d.** Since sets are mutable, a set cannot contain another set. For this, one has to use tuples (but then being careful about the order) or immutable sets `frozenset`:

```python
def aretes(graph):
    # set of edges
    out = set()
    # loops on all edges
    for s in graph:
        for a in graph[s]:
            out.add(frozenset((s,a)))
    return out
```

**e.** The idea is to add the optional variable path to the recursive function in order to keep in memory the first part of the path already explored during the recursive calls:

```python
def find_path(graph, start, end, path=[]):
    # add to path without modifying the argument
    path = path + [start]
    # exit is path terminated
    if start == end:
        return path
    # exit is start is isolated
    if start not in graph:
        return None
    # loop on the adjacent vertices not already in the path
    for node in graph[start]:
        if node not in path:
            # search for a new path from here
            newpath = find_path(graph, node, end, path)
            # exit if the new path is OK
            if newpath: return newpath
    return None
```

For example:

```python
find_path(Ga,1,6)
find_path(Gc,0,7)
```

**Remark:** If we replace the third line `path = path + [start]` with `path += [start]` or with `path.append(start)`, the code doesn't work because it changes the function argument itself. For example, test the function:

```python
def append_to(element, to=[]):
    to.append(element)
    return to
```

by executing:

```
print(append_to(1))
print(append_to(2))
print(append_to(2,[]))
```

The conclusion is to never modify a mutable optional parameter (unless you do it on purpose to make a cache, for example). The previous function should have been written:

```
def append_to(element, to=None):
    if to is None:
        to = []
    to.append(element)
    return to
```

**f.** The idea is almost identical to that of the previous function:

```
def find_paths(graph, start, end, path=[]):
    # exit is path terminated
    if start == end:
        return [path+[start]]
    # exit is start is isolated
    if start not in graph:
        return []
    # list of paths
    paths = []
    # loop on the adjacent vertices not already in the path
    for node in graph[start]:
        if node not in path and node!=start:
            # search for a new path from here
            newpaths = find_paths(graph, node, end,
            ↳   path+[start])
            paths += newpaths
    return paths
```

For example:

```
find_paths(Ga,1,6)
find_paths(Gc,0,7)
```

## SOLUTION 7.2   TRIANGLES IN A GRAPH

**a.** In order not to count each set of three elements twice, we have to classify them. Thus, the set of subsets of cardinal three of $X = \{1, 2, \dots, n\}$ is:

$$\{(i, j, k) : 1 \leq i < j < k \leq n\},$$

and its cardinal is:

$$\sum_{i=1}^{n}\sum_{j=i+1}^{n}\sum_{k=j+1}^{n}1 = \sum_{i=1}^{n}\sum_{j=i+1}^{n}(n-j) = \sum_{i=1}^{n}\frac{(n-i)(n-i-1)}{2} = \frac{n(n-1)(n-2)}{6}.$$

It is also possible to see this cardinal geometrically as the sixth of a cube by paying attention to the equal elements. This is also the binomial coefficient:

$$\binom{n}{3} = \frac{n!}{(n-3)!\,3!} = \frac{n(n-1)(n-2)}{6}.$$

Even for a medium size graph, this number of possible triangles is quite large, thus a better approach as to be used.

**b.** For each edge $\{i, j\}$, we test the set of edges $\{i, k\}$, starting from $i$ and see if $\{j, k\}$ is an edge:

```python
def triangles(graph):
    # set of triangles
    out = set()
    # loop on the edges
    for i in graph:
        for j in graph[i]:
            # loop on the flower of i
            for k in graph[i]:
                # if this is a triangle
                if k in graph[j]:
                    out.add(frozenset((i,j,k)))
    return out
```

To test:

```python
triangles(Ga), triangles(Gb), triangles(Gc)
```

**c.** In order to make sure that even graphs with non-integer keys can be processed, we define a dictionary of correspondence between the vertices and the set $\{0, 1, \dots, n-1\}$:

```python
import numpy as np
def adjacency(graph):
    # dimension of the graph
    n = len(graph)
    # adjacency matrix
    A = np.zeros((n,n), dtype=int)
    # define the correspondence dictionary vertex -> index
    ↳   between 0 and n
    dico = {i: idx for idx,i in enumerate(graph)}
    for i in graph:
        for j in graph[i]:
```

```
            A[dico[i],dico[j]] = 1
    return A
```

To test:

```
adjacency(Ga), adjacency(Gb), adjacency(Gc)
```

**d.** A very simple way is to use dictionary comprehension:

```
def graph(A): # A is supposed to be a square matrix
    n = len(A)
    return {i: {j for j in range(n) if A[i][j]==1} for i in
    ↳  range(n)}
```

Another way is to use the where function of NumPy which allows to return the indices (in tuple of length 1) of the entries with a 1:

```
def graph(A):
    graph = dict()
    for i,line in enumerate(A):
        # indices with 1 one this row
        indices, = np.where(line)
        # add the key and corresponding indices as value
        graph[i] = set(indices)
    return graph
```

To test:

```
graph(adjacency(Ga)) == Ga, \
graph(adjacency(Gb)) == Gb, \
graph(adjacency(Gc)) == Gc
```

Note that graphe(adjacence(Ga)) is not equal to Ga because the numbering starts once at 0 and once at 1.

**e.** An edge $\{i, j\}$ is a member of a triangle if and only if $A_{ij} \neq 0$ and $B_{ij} \neq 0$. Indeed,

$$B_{ij} = \sum_{k=1}^{n} A_{ik}A_{kj},$$

and therefore if $B_{ij} \neq 0$, then there exists at least one $k$ such that $A_{ik}A_{kj} = 1$ so $\{i, j, k\}$ is a triangle. Thus:

```
def triangles2(A):
    # set of triangles
    out = set()
    # matrix product
    B = A @ A
    # matrix defined by True if A_ij \neq 0 and B_ij \neq 0
    cond = (A !=0) & (B !=0)
```

```
    # indices of edges from which a triangle exists
    lst_i,lst_j = np.where(cond)
    for i,j in zip(lst_i,lst_j):
        # indices k such that A_ik=1 and A_jk=1
        indices, = np.where( (A[i]==1) & (A[j]==1) )
        # al these indices are triangles
        for k in indices:
            out.add(frozenset((i,j,k)))
    return out
```

To test with a random matrix:

```
# random matrix of 0 and 1
A = np.random.binomial(1,0.5,(100,100))
# symmetrize the matrix
A = A & A.T
# test
triangles2(A) == triangles(graph(A))
```

**f.** The number of distinct paths of length $n$ between $i$ and $j$ is given by $C_{ij}$ where $C = A^n$. Thus, the trace of $A^3$ represents the number of triangles present in the graph with their multiplicity. Since the triangles can start with one of three different vertices and in two different directions, the number of distinct triangles is $\frac{\mathrm{tr}(A^3)}{6}$:

```
def nb_triangles(A):
    # calculate A^3 in the matrix sense
    B = np.linalg.matrix_power(A,3)
    return np.trace(B)//6
```

To test:

```
nb_triangles(adjacency(Ga)), nb_triangles(adjacency(Gb)),
 ↳  nb_triangles(adjacency(Gc))
```

## SOLUTION 7.3   MODULE NETWORKX (!!)

**b.** We choose to analyze a graph representing the emails exchanged within a university over a given period of time, as described at: https://snap.stanford.ed u/data/email-EuAll.html. For a given email, the nodes are the email addresses. An edge is created between two nodes $i$ and $j$ if $i$ has sent at least one email to $j$. Here, the graph is considered as undirected.
The first thing is to load the NetworkX module (version 2) and Matplotlib:

```
import networkx as nx
import matplotlib.pyplot as plt
```

Then, we download the graph and open it with NetworkX. The `read_edgelist` function allows to define a graph from the list of edges:

```python
import urllib.request, gzip, io
# url to download
url = "https://snap.stanford.edu/data/email-EuAll.txt.gz"
# download the gz file
file = urllib.request.urlopen(url)
# extract the compressed gz
dat = gzip.GzipFile(fileobj=io.BytesIO(file.read()))
# import the list of edges
g = nx.read_edgelist(dat)
```

Once the graph is defined, we can, for example, determine the number of nodes:

```python
g.number_of_nodes()
```

or its density:

```python
nx.density(g)
```

To get an idea of how the graph looks, we plot the degree distribution of the nodes as in Figure 7.1:

```python
degree = dict(g.degree())
degree = sorted(degree.values(), reverse=True)
plt.figure(figsize=(8,5))
plt.title("Degree distribution")
plt.xlabel("Rank")
plt.ylabel("Degree")
plt.loglog(degree)
```

To find out the number of related components:

```python
nx.number_connected_components(g)
```

To have more reasonable sizes of graphs, we choose the related components that have between 50 and 100 elements:

```python
subgraphs = []
for s in nx.connected_components(g):
    if 50 <= len(s) <= 100:
        h = nx.convert_node_labels_to_integers(g.subgraph(s))
        subgraphs.append(h)
```

and represent them graphically as in Figure 7.2:

```python
fig = plt.figure(figsize=(12,5))
fig.suptitle(r'Connected components with 50 to 100 elements')
for i,s in enumerate(subgraphs):
```

```
    sub = fig.add_subplot(1,2,i+1)
    nx.draw(s, with_labels=True)
```

To determine the diameter of the second:

```
nx.diameter(subgraphs[1])
```

It is possible to detect the communities as represented in Figure 7.3:

```
from networkx.algorithms import community
fig = plt.figure(figsize=(12,5))
fig.suptitle(r'Communities detection')
for i,s in enumerate(subgraphs):
    partition = community.greedy_modularity_communities(s)
    color = [0]*s.number_of_nodes()
    for c,p in enumerate(partition):
        for j in p:
            color[j] = c
    sub = fig.add_subplot(1,2,i+1)
    nx.draw(s, node_color=color, with_labels=True)
```

We can also calculate the *PageRank* of each node as shown in Figure 7.4:

```
fig = plt.figure(figsize=(12,5))
fig.suptitle(r'PageRank')
for i,s in enumerate(subgraphs):
    pk = nx.pagerank(s)
    sub = fig.add_subplot(1,2,i+1)
    nx.draw(s, node_color=list(pk.values()),
        with_labels=True)
```

Finally, we select the subgraph whose nodes have a degree between 50 and 100:

```
h = g.subgraph([x for x in g.nodes() if 50 <= g.degree(x)
    <=100])
len(h),nx.number_connected_components(h)
```

and the main related component is selected:

```
hc = h.subgraph(max(nx.connected_components(h), key=len))
nx.draw(hc)
```

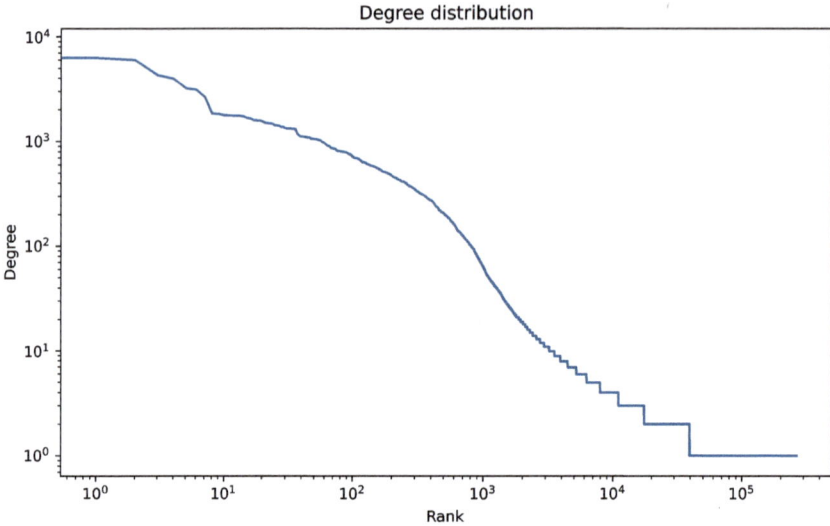

Figure 7.1    Distribution of the degrees of the nodes in the graph.

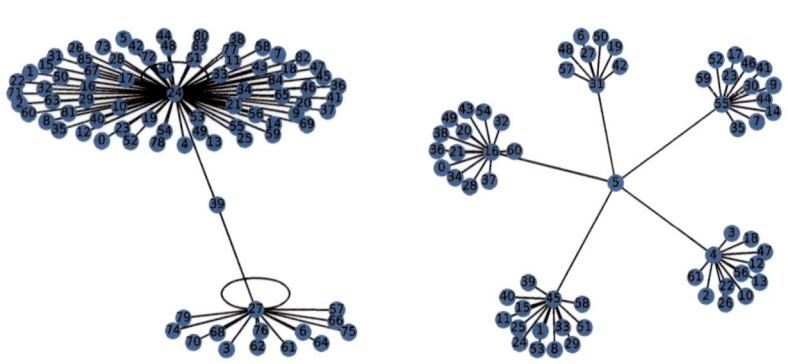

Figure 7.2    Graphical representation of the two connected components with be-tween 50 and 100 nodes in the graph.

Communities detection

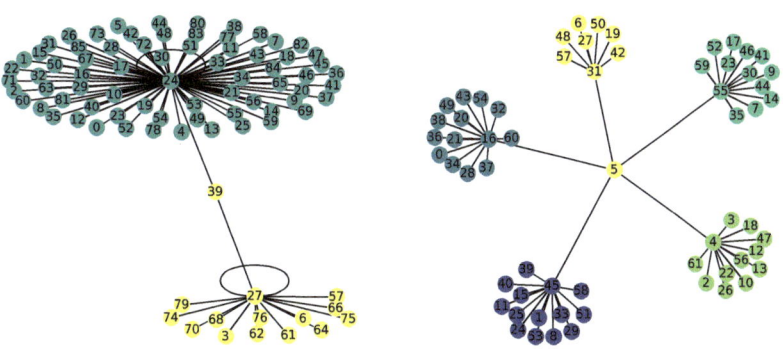

Figure 7.3    Representation of the communities composing the two sub-graphs.

PageRank

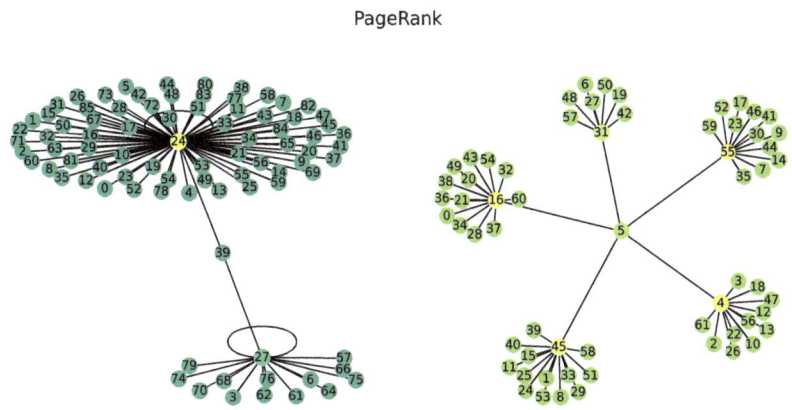

Figure 7.4    Determine the *PageRank* of each node for the two subgraphs.

# CHAPTER 8

# Symbolic Calculation

As a general purpose language, Python does not include by default some mathematical concepts. An example already seen concerns vectors and numerical matrices which are implemented in the NumPy module. The goal here is to introduce the SymPy module which allows to do symbolic calculation.

For example, the number $\sqrt{8}$ is represented by default in Python as a float. The advantage of SymPy is that $\sqrt{8}$ is kept as a root and even automatically simplified:

```
import sympy as sp
sp.init_printing()
sp.sqrt(8)
```

Note that the second instruction is not necessary, but allows to present the results in a more elegant way in Jupyter Lab. SymPy documentation is available at the address: https://docs.sympy.org/.

Concepts covered

- symbols and symbolic expressions

- simplification

- infinitesimal analysis (limit, derivation, integration, series)

- computer-assisted proof

- pathological function

- Green's function

- spherical coordinates

DOI: 10.1201/9781003565451-8

# EXERCISES

## EXERCISE 8.1   INTRODUCTION TO SYMPY

Before you can use symbolic variables, you have to declare them as symbols:

```
x = sp.Symbol("x") # define the symbol x
y = sp.Symbol("y", real=True) # define a real varaible y
e = sp.Symbol(r"\varepsilon", real=True, positive=True) #
   define a positive variable
```

Then, it is possible to perform operations between symbols:

```
x + 2*y + e/4 + x**2 + 3*x + 2*y
```

Most of the mathematical functions are implemented symbolically in SymPy and it is also possible to simplify them:

```
expr = sp.cos(x)**2 + sp.sin(x)**2 + (y**3 + y**2 - y -
   1)/(y**2 + 2*y + 1) + sp.exp(-e)
sp.simplify(expr)
```

Finally, it is possible to make substitutions:

```
expr.subs(x,y) # substitute x by y
expr.subs({y:x, e:y}) # substitute y by x and e by y
```

then, for example, simplify the expression and plot its graph as a function of x as in Figure 8.1:

```
f = sp.simplify(expr.subs({y:x, e:y}))
sp.plot(f,(x,-2,6), title=f"Plot of ${sp.latex(f)}$")
```

**a.** Read the documentation for the solve function and use it to calculate the roots of a general polynomial of degree two, then of degree three.
*Hint: The documentation on solving algebraic equations is available at the address:* https://docs.sympy.org/latest/modules/solvers/solvers.html#algebraic-equations.

**b.** Read the documentation for the functions evalf and N to evaluate numerically the expression $\frac{\pi^2}{4}$.
*Hint: The documentation on the numerical evaluation is available at the address:* https://docs.sympy.org/latest/modules/evalf.html.

**c.** Read the documentation for the Rational function and numerically evaluate the rational number $\frac{43609}{999}$ to 50 decimal places.

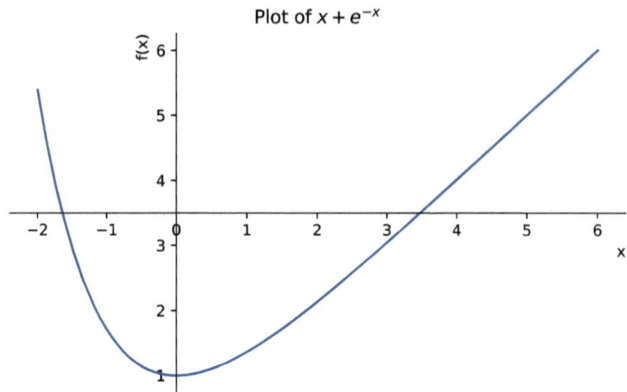

Figure 8.1    Plot of a SymPy function.

**d.** Determine the real and imaginary part of the expression:

$$\left(\frac{1+i\sqrt{3}}{1+i}\right)^{20}.$$

*Hint: See the documentation at the address:* `https://docs.sympy.org/latest/modules/functions/elementary.html`.

**e.** Read the documentation for the function `diff` and calculate the derivative of $xe^{x^{x^x}}$ with respect to $x$.

*Hint: The documentation on derivatives is available at the address:* `https://docs.sympy.org/latest/tutorial/calculus.html#derivatives`.

**f.** Read the documentation of the function `integrate` and calculate the following integrals:

$$I_1 = \int x^5 \sin(x)\,dx, \qquad\qquad I_2 = \int_0^\infty \sin(x^2)\,dx.$$

**g.** Calculate with SymPy the following limits:

$$L_1 = \lim_{x\to 0}\frac{\sin(x)}{x}, \qquad L_2 = \lim_{x\to 0}\sin\left(\frac{1}{x}\right), \qquad L_3 = \lim_{x\to\infty}\frac{5x^2+3x+2y}{y(x-4)(x-y)}.$$

**h.** Compute the series expansion of $\tan(x)$ at $x = 0$ to order 10 and the asymptotic expansion of $\left(1 + \frac{1}{n}\right)^n$ for $n \to \infty$ to order 5.

**i.** Determine the eigenvalues of the matrix:

$$\begin{pmatrix} 1 & a & 0 \\ a & 2 & a \\ 0 & a & 3 \end{pmatrix}.$$

*Hint: The documentation on symbolic matrices is available at the address:* `https://docs.sympy.org/latest/tutorial/matrices.html`.

## EXERCISE 8.2   APPLICATIONS

The goal is to use SymPy to solve symbolically different mathematical problems by calculating the least possible things by hand.

**a.** Determine the number of zeros in the integer 123!.

**b.** Determine the ratio between the height and radius of a cylinder so as to minimize its area at a fixed volume.

**c.** For $x, y \in \mathbb{R}$ such that $xy < 1$, show that

$$\arctan(x) + \arctan(y) = \arctan\left(\frac{x+y}{1-xy}\right).$$

*Hint: Derive the equation with respect to x and justify.*

**d.** Prove the following formula due to Gauss:

$$\frac{\pi}{4} = 12 \arctan\left(\frac{1}{38}\right) + 20 \arctan\left(\frac{1}{57}\right) + 7 \arctan\left(\frac{1}{239}\right) + 24 \arctan\left(\frac{1}{268}\right).$$

It is imperative to use SymPy, the original demonstration of Gauss being 25 pages long, see pages 477 to 502 of the second volume of his complete works available at the address: `https://gallica.bnf.fr/ark:/12148/bpt6k99402s`.
*Hint: Apply the tangent function on each side of the equation and simplify. The documentation on the different simplification functions is available at the address:* `https://docs.sympy.org/latest/tutorial/simplification.html`.

**e.** Determine the volume of the region:

$$\{(x, y, z) \in \mathbb{R}^3 : x^2 + y^2 < z < 2x^2 + 4xy + 6y^2, |y| < 5, |x| < 4\}.$$

**f.** Determine the expression of the real Fourier coefficients of the $2\pi$-periodic function $f$ defined by $f(x) = |\sin(x)|$.

## EXERCISE 8.3   CONJECTURE DUE TO EULER

Euler conjectured in 1769 that at least $k$ powers of strictly positive integers are necessary for the sum to be itself a $k$ power. In other words, if $n \geq 2, k \geq 1$, $a_1, a_2, ..., a_n \geq 1$ and $b \geq 1$ are integers such that:

$$\sum_{i=1}^{n} (a_i)^k = b^k$$

then necessarily $n \geq k$. This conjecture was disproved in 1966 by Lander & Parkin (doi:10.1090/S0002-9904-1966-11654-3) in what appears to be the shortest mathematical paper ever written with a counterexample for $k = 5$:

$$27^5 + 84^5 + 110^5 + 133^5 = 144^5.$$

The goal is to show that this counterexample is the simplest possible, in the sense that it is the only counterexample with $k \leq 5$ and $b \leq 144$.

**a.** Show that Euler's conjecture is true for $k = 1$ and $k = 2$.

**b.** Check with Python the above counterexample.

**c.** Write a function powers(bmax,k) that returns the set (type set) of all integers from 1 to bmax raised to the power k.

**d.** Check that there is no counterexample with $k = 3$ and $b \leq 144$.

**e.** Write a function combinations(lst,n) that for a list of integers lst and an integer n returns the list of all combinations of n integers in lst in ascending order. For example, combinations([1,2,3,4],2) should return [(1,1), (1,2), (1,3), (1,4), (2,2), (2,3), (2,4), (3,3), (3,4), (4,4)].
*Hint: Use a recursive function on n.*

**f.** Write a function test(bmax,n,k) that for three given integers bmax, n, and k, iterates over all combinations of n integers returned by combinations and checks whether the sum of these n integers raised to the power k is an integer present in the list powers(bmax,k). Use this function to check that there is no counterexample to Euler's conjecture for $k = 4$ and $b \leq 144$. Depending on the power of your computer, it is possible to choose also $k = 5$ and thus to check that the counterexample of the introduction is indeed the simplest one.
For $k = 5$, the previous method of iterating over all combinations is rather slow. A faster method is to observe that the set of sums of type:

$$(a_1)^5 + (a_2)^5 + (a_3)^5 + (a_4)^5,$$

can be written as $S_1 + S_2$, where $S_1$ and $S_2$ are sums of two integers to the power 5.

**g.** Write a function sum2(bmax,k) which returns a dictionary having for keys the sums $(a_1)^k + (a_2)^k$ with the associated value $(a_1, a_2)$ for $0 \leq a_1 \leq a_2 \leq$ bmax. We will take care to remove the trivial element zero from the dictionary. When building the dictionary, make sure that it is uniquely defined in the sense that there is no other possible value for an existing key. Test sum2(300,5) and sum2(300,3).

**h.** Use the dictionary constructed earlier to determine the set of counterexamples for $k = 5$ and $b \leq 300$ by iterating over all elements of powers(bmax,5) and sum2(bmax,5).
*Hint: An optimal implementation takes at most a few seconds to run.*

## EXERCISE 8.4   PATHOLOGICAL FUNCTION

The goal is to construct a function that visually looks regular, but in fact is not. Let the function $f : \mathbb{R} \to \mathbb{R}$ be defined by:

$$f(x) = \sum_{k=1}^{\infty} \frac{\sin(k^2 x)}{k^5} .$$

Since the series converges absolutely, the function $f$ is well defined.

**a.** With the help of SymPy calculate the function $g : \mathbb{R}$ defined by keeping the first hundred terms of the series:

$$g(x) = \sum_{k=1}^{100} \frac{\sin(k^2 x)}{k^5} ,$$

and plot the function g.

**b.** Estimate by hand the error between the functions $f$ and $g$ in absolute value.

**c.** Calculate the first derivative and the second derivative of g and plot these two derivatives. What can you conclude?

**d.** Explain mathematically what is going on.

## EXERCISE 8.5   GREEN'S FUNCTION OF THE LAPLACIAN (!)

The goal of this exercise is to compute fully automatically the Green's function of the Laplacian in $\mathbb{R}^3$, i.e., the solution satisfying:

$$\Delta G(\mathbf{x}) = \delta(\mathbf{x}),$$

in $\mathbb{R}^3$, where $\delta(\mathbf{x})$ is the Dirac distribution.
For this, we introduce the spherical coordinates $\mathbf{x}' = (r, \theta, \varphi)$ with $r > 0, 0 \leq \theta \leq \pi$ and $0 \leq \varphi < 2\pi$ characterized by:

$$x_1 = r \cos \varphi \sin \theta$$
$$x_2 = r \sin \varphi \sin \theta$$
$$x_3 = r \cos \theta .$$

**a.** Define a function to_spherical(expr) to convert an expression given in Cartesian coordinates to spherical coordinates.

**b.** Define a function to_cartesian(expr) allowing to convert into cartesian co-ordinates an expression given in spherical coordinates. For simplicity, we can only deal with the case of an expression expr invoking the variables $r$ and $\theta$ but not $\varphi$.

**c.** Calculate the scale factors of the spherical coordinates:

$$h_i = \left\| \frac{\partial \mathbf{x}}{\partial x_i'} \right\| .$$

**d.** Define a function `gradient(f)` allowing to calculate the gradient of a function $f : \mathbb{R}^3 \to \mathbb{R}$ in spherical coordinates:

$$\nabla f = \left( \frac{1}{h_i} \frac{\partial f}{\partial x_i'} \right)_{i=1}^{3}.$$

**e.** Do the same to define the Laplacian in spherical coordinates:

$$\Delta f = \sum_{i=1}^{3} \frac{1}{J} \frac{\partial}{\partial x_i'} \left( \frac{J}{h_i^2} \frac{\partial f}{\partial x_i'} \right) \quad \text{where} \quad J = \prod_{i=1}^{3} h_i.$$

**f.** Find the radial solutions (*i.e.*, depending only on the variable $r$) of the equation $\Delta G = 0$ in $\mathbb{R}^3 \setminus \{0\}$.
*Hint: Look at the documentation of the function `dsolve` to solve a differential equation.*

**g.** Determine the equations that the integration constants must satisfy for the above solution to satisfy in Cartesian coordinates:

$$\lim_{|x| \to \infty} G(x) = 0 \quad \text{and} \quad \Delta G(x) = \delta(x).$$

*Hint: We must transform the two conditions into spherical coordinates. The first condition is expressed in spherical coordinates by:*

$$\lim_{r \to \infty} G(r) = 0,$$

*and the second is equivalent to:*

$$\lim_{r \to 0} \int_0^{\pi} \int_0^{2\pi} \left( \nabla G(r) \cdot e_r \right) J(r, \theta, \varphi) \, \mathrm{d}\varphi \mathrm{d}\theta = 1.$$

**h.** Solve the equations on the integration constants and substitute in the radial solution of $\Delta G = 0$ to obtain the expression of the Green's function of the Laplacian in spherical coordinates. Finally, determine the Green's function $G$ of the Laplacian in $\mathbb{R}^3$ in Cartesian coordinates.

**i.** ‼ Let $g : \mathbb{R}^3 \to \mathbb{R}$ be a smooth function with compact support invariant by rotations along the vertical axis. Determine the asymptotic behavior at large distances of the solution of the equation:

$$\Delta f(x) = g(x)$$

up to order two, *i.e.*, the terms decreasing as $|x|^{-1}$ and as $|x|^{-2}$.
*Hint: The solution is given by Green's formula:*

$$f(x) = \int_{\mathbb{R}^3} G(x - x_0) g(x_0) \, \mathrm{d}^3 x_0 = \int_{B_R} G(x - x_0) g(x_0) \, \mathrm{d}^3 x_0,$$

*where R is chosen so that the support of g is contained in $B_R$. To compute the asymptotic expansion of this integral, the first step is to convert $G(\boldsymbol{x} - \boldsymbol{x}_0)$ into spherical coordinates for $\boldsymbol{x}$ and $\boldsymbol{x}_0$. Since g is invariant by rotations along the vertical axis, then in spherical coordinates g is independent of $\varphi$ and is given by $g(r, \theta)$. The second step is to transform the integral in spherical coordinates with a triple integral on $r_0$, $\theta_0$, and $\varphi_0$. The third step is to compute the asymptotic development of the integrand when $r \to \infty$, up to order two. Finally, since $r_0$, $\theta_0$, and $\varphi_0$ are bounded, the integration then commutes with the asymptotic expansion and the final result is given by the individual integration of the two terms of the asymptotic expansion.*

# SOLUTIONS

## SOLUTION 8.1   INTRODUCTION TO SYMPY

**a.** Just define the symbols and then solve the equation with the `solve` function:

```
a,b,c,d = sp.symbols("a b c d")
sp.solve(a*x**2+b*x+c,x)
sp.solve(a*x**3+b*x**2+c*x+d,x)
```

**b.** Your choice:

```
(sp.pi**2/4).evalf()
sp.N(sp.pi**2/4)
```

**c.** The rational number is defined using the `Rational` function and then evaluated to 50 decimal places:

```
frac = sp.Rational(43609, 999)
frac.evalf(50)
```

**d.** There is two possibilities:

```
expr = sp.simplify(((1+sp.I*sp.sqrt(3))/(1+sp.I))**20)
(sp.re(expr),sp.im(expr))
expr.as_real_imag()
```

**e.** The function `diff` allows to calculate symbolically the derivative of a symbolic expression:

```
sp.diff(x*sp.exp(x**x**x),x)
```

**f.** It is possible to calculate definite or indefinite integrals:

```
I1 = sp.integrate(x**5*sp.sin(x),x)
I2 = sp.integrate(sp.sin(x**2), (x,0,sp.oo))
```

Note that infinity is represented in SymPy by the symbol oo.

**g.** The limits of expressions requested:

```
L1 = sp.limit(sp.sin(x)/x, x, 0)
L2 = sp.limit(sp.sin(1/x), x, 0)
L3 = sp.limit((5*x**2+3*x+2*y)/(y*(x-4)*(x-y)),x,sp.oo)
```

**h.** The limited or asymptotic developments are computed with the `series` function:

```
S1 = sp.series(sp.tan(x), x, 0, 10)
S2 = sp.series((1+1/x)**x, x, sp.oo, 5)
```

**i.** First, we have to construct the symbolic matrix and then calculate its eigenvalues:

```
M = sp.Matrix([[1,a,0],[a,2,a],[0,a,3]])
M.eigenvals()
```

Note that the multiplicity of eigenvalues is also returned.

## SOLUTION 8.2  APPLICATIONS

**a.** One way is to evaluate 123! and then convert it to a string to count the number of zeros:

```
fact = sp.factorial(123)
str(fact).count('0')
```

and discover that there are 42.

**b.** The variables defining the problem and the equations:

```
r = sp.Symbol("r", real=True, positive=True) # radius
h = sp.Symbol("h", real=True, positive=True) # height
V = sp.Symbol("V", real=True, positive=True) # volume
volume = sp.pi*r**2*h # volume of the cylinder
surface = 2*sp.pi*r**2 + 2*sp.pi*r*h # area of the cylinder
```

The first step is to remove h from the first equation and replace it with V in the second:

```
solh = sp.solve(volume-V,h)[0]
S = surface.subs(h,solh)
```

Finally, we have to minimize S with respect to r and solve for r:

```
rmin = sp.solve(sp.diff(S,r),r)[0]
sp.solve(rmin.subs(V,volume)-r,r)
```

Thus, the cylinder of minimal area with fixed volume verifies $r = \frac{h}{2}$.

**c.** The function $f$ formed by the difference of the two sides of the equation:

$$f(x,y) = \arctan(x) + \arctan(y) - \arctan\left(\frac{x+y}{1-xy}\right),$$

is defined for all $x, y \in \mathbb{R}$ such that $xy \neq 1$. By taking the partial derivative of $f$ with respect to $x$:

```
x = sp.Symbol("x", real=True)
y = sp.Symbol("y", real=True)
f = sp.atan(x) + sp.atan(y) - sp.atan((x+y)/(1-x*y))
sp.simplify(sp.diff(f,x))
```

we obtain that the latter is zero on the domain of definition of $f$. For any fixed $y \in \mathbb{R}$, since $f(0, y) = 0$ and $f(x, y)$ is continuously differentiable in $x$ for $xy < 1$ and of zero derivative, then $f(x, y) = 0$ for $xy < 1$, which ends the proof.

**d.** The first step is to define the right-hand side of the equation, denoted $\Theta$:

```
T = 12*sp.atan(sp.Rational(1,38))   +
    20*sp.atan(sp.Rational(1,57)) \
  + 7*sp.atan(sp.Rational(1,239)) +
      24*sp.atan(sp.Rational(1,268))
```

To keep the symbolic character of the expression, it is imperative to use the Rational function, otherwise $1/38$ is directly evaluated as a floating-point number by Python. The second step is to apply the tangent function to $\Theta$ and to simplify with expand_trig:

```
sp.expand_trig(sp.tan(T))
```

which shows that $\tan(\Theta) = 1$. This last equation has an infinity number of solutions, but verifying that $0 < \Theta < \frac{\pi}{2}$:

```
0 < T < sp.pi/2
```

then it shows that $\Theta = \arctan(1) = \frac{\pi}{4}$ since the tangent function is bijective on $]0, \frac{\pi}{2}[$.

**e.** One way to do this is to integrate the constant function 1 over the specified domain:

```
x,y,z = sp.symbols("x y z")
sp.integrate(1,(z,x**2+y**2,2*x**2+4*x*y+6*y**2),(y,-|
   5,5),(x,-4,4))
```

**f.** We define the function $f$, then the symbol corresponding to $n \geq 1$ allowing to calculate the integrals defining the Fourier coefficients:

```
x = sp.Symbol("x", real=True)
n = sp.Symbol("n", integer=True, positive=True)
f = abs(sp.sin(x))
a0 = sp.integrate(f,(x,-sp.pi,sp.pi))/(2*sp.pi)
an = sp.integrate(f*sp.cos(n*x),(x,-sp.pi,sp.pi))/sp.pi
bn = sp.integrate(f*sp.sin(n*x),(x,-sp.pi,sp.pi))/sp.pi
sp.simplify((a0,an,bn))
```

## SOLUTION 8.3   CONJECTURE DUE TO EULER

**a.** Since by hypothesis $n \geq 2$, then necessarily $n \geq k$.

**b.** List comprehension allows the sum of the fifth powers:

```
sum([i**5 for i in [27,84,110,133]]) == 144**5
```

**c.** The easiest way is to use set comprehension:

```
def powers(bmax,k):
    return {i**k for i in range(1,bmax+1)}
```

**d.** If we want to find a counterexample for $k = 3$, then necessarily $n = 2$. So, we just have to test all combinations of two integers $a_1$ and $a_2$:

```
def check3(bmax):
    out = set()
    integers3 = powers(bmax,3)
    for a1 in range(1,bmax+1):
        for a2 in range(a1,bmax+1):
            s = a1**3 + a2**3
            if s in integers3:
                out.add(s)
    return out
```

which gives an empty set for $b \leq 144$:

```
check3(144)
```

**e.** The idea is to make a recursive function adding the new cases to the output list:

```
def combinations(lst,n):
    if n==1:
        return [(i,) for i in lst]
    else:
        out = []
        for l in combinations(lst,n-1):
            for i in lst:
                if i >= l[-1]: out.append(l+(i,))
        return out
```

for example:

```
combinations([1,2,3,4],2)
```

**f.** Just iterate on the combinations:

```python
def test(bmax,n,k):
    out = set()
    p = powers(bmax,k)
    for c in combinations(range(1,bmax+1),n):
        s = sum([i**k for i in c])
        if s in p:
            out.add(c)
    return out
```

For $k = 3, 4$, we do not find a solution:

```python
test(144,2,3) | test(144,2,4) | test(144,3,4)
```

For $k = 5$, we find the expected solution:

```python
test(144,2,5) | test(144,3,5) | test(144,4,5)
```

Note that it is possible to improve the performance of this algorithm a bit by doing the combinations on powers `powers(bmax,k)` rather than on integers. This algorithm stores all the combinations in memory, but the use of the `combinations_` with_replacement function of `itertools` allows to get rid of this problem.

**g.** The idea is to iterate over all the values of $a_1 \leq a_2$ and to test if the key does not already exist:

```python
def sum2(bmax,k):
    out = dict()
    for i in range(0,bmax+1):
        for j in range(i,bmax+1):
            s = i**k + j**k
            # this should never happen, otherwise the
            ↵  dictionary is not uniquely defined
            if s in out: raise ValueError("Dictionary not
            ↵  unique")
            out[s] = (i,j)
    del out[0] # remove 0:(0,0) from the dictionary
    return out
```

For $k = 5$, the dictionary seems to be uniquely defined:

```python
sum2(300,5)
```

whereas for $k = 3$, this is not the case:

```python
sum2(300,3)
```

Note that the unique dictionary definition for k=5 and any value of `bmax` is an open mathematical problem.

**h.** It is important to iterate only once on sum2(bmax,5) and to test the membership of the difference to sum2(bmax,5):

```python
def test5(bmax):
    # list of counter-examples
    out = set()
    # dictionary of the sum of two 5th power
    s2 = sum2(bmax,5)
    # loop on the 5th powers
    for p in powers(bmax,5):
        for s in s2:
            if p - s in s2:
                # sort the counter-example values
                l = sorted(s2[s] + s2[p-s])
                out.add(tuple(l))
    return out
```

This allows us to discover a second counter-example:

```python
test5(300)
```

given by:

$$54^5 + 168^5 + 220^5 + 266^5 = 288^5.$$

Note that it is possible to optimize this code by sorting the result of sum2(bmax,5) beforehand in order to exit the loop when it is no more useful:

```python
def test5opt(bmax):
    # list of counter-examples
    out = set()
    # dictionary of the sum of two 5th power
    s2 = sum2(bmax,5)
    s2sorted = sorted(s2)
    # loop on the 5th powers
    for p in powers(bmax,5):
        for s in s2sorted:
            if p <= s: break
            if p - s in s2:
                # sort the counter-example values
                l = sorted(s2[s] + s2[p-s])
                out.add(tuple(l))
    return out
```

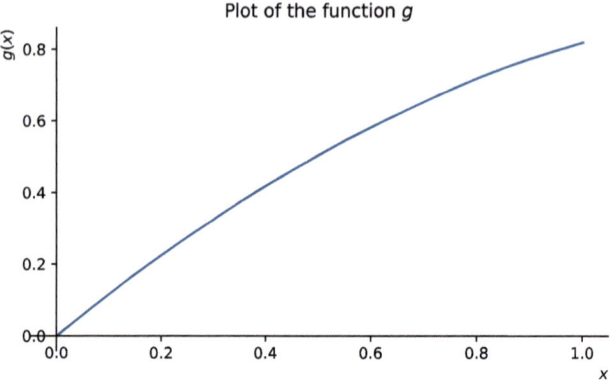

Figure 8.2    Plot of the function g.

## SOLUTION 8.4    PATHOLOGICAL FUNCTION

**a.** We define the variables, the term of the series, and then the sum of the first hundred terms:

```
x = sp.Symbol("x")
k = sp.Symbol("k")
g = sp.summation(sp.sin(k**2*x)/k**5, (k,1,100))
```

Note that it is important to use the summation function and not Sum which keeps the sum in symbolic form. To represent the function g graphically as in Figure 8.2:

```
sp.plot(g,(x,0,1), ylabel=r"$g(x)$")
```

**b.** The error is bounded by:

$$|f(x) - g(x)| \leq \left| \sum_{k=101}^{\infty} \frac{\sin(k^2 x)}{k^5} \right| \leq \sum_{k=101}^{\infty} \frac{1}{k^5} \leq \int_{100}^{\infty} \frac{1}{k^5} dk \leq 2.5 \times 10^{-9},$$

so g seems to be a very good approximation of f.

**c.** It suffices to take the derivatives with SymPy and plot them to get Figure 8.3:

```
gp = sp.diff(g,x)
gpp = sp.diff(gp,x)
sp.plot(gp,(x,0,1), ylabel=r"$g^{\prime}(x)$")
sp.plot(gpp,(x,0,1), ylabel=r"$g^{\prime\prime}(x)$")
```

The conclusion is that g′ seems to be continuous while g″ does not.

Plot of the derivatives of the function g

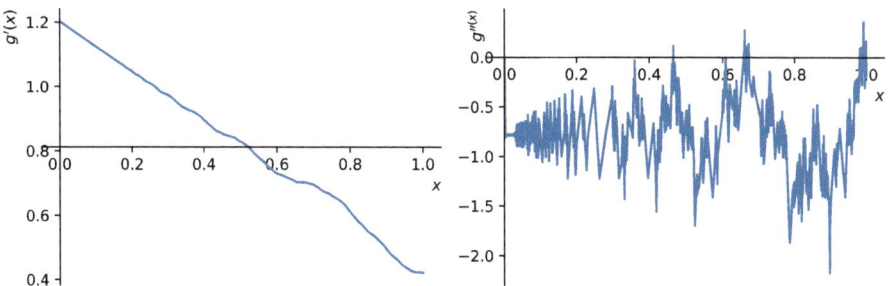

Figure 8.3    Plot the derivatives of the function g. The first derivative still seems regular while the second derivative does not seem continuous.

**d.** Since the series defining $f$ is uniformly convergent, then $f$ is continuous. By formally deriving term by term:

$$f'(x) = \sum_{k=1}^{\infty} \frac{\cos(k^2 x)}{k^3},$$

and since this series converges uniformly, then $f$ is continuously derivable and the derivative is given by the above series. By deriving again formally:

$$f''(x) = \sum_{k=1}^{\infty} \frac{-\sin(k^2 x)}{k}.$$

However, this series is not uniformly convergent because it is not even convergent for all values of $x$ (take for example $x = \frac{\pi}{2}$). Thus, $f'$ is not differentiable.

## SOLUTION 8.5   GREEN'S FUNCTION OF THE LAPLACIAN (!)

**a.** First, we need to define the symbols $x, y, z$ of the Cartesian coordinates:

```
x, y, z = sp.symbols("x y z", real=True)
xyz = [x,y,z]
```

then the symbols of the spherical coordinates:

```
r, t, p = sp.symbols(r"r \theta \varphi", real=True,
 ↪  positive=True)
rtp = [r,t,p]
```

By defining the spherical coordinates by a dictionary:

```
coords = {x: r*sp.cos(p)*sp.sin(t), y: r*sp.sin(p)*sp.sin(t),
    z: r*sp.cos(t)}
```

it is very easy to write a function to convert an expression into spherical coordinates:

```
def to_spherical(expr):
    return sp.simplify(expr.subs(coords))
```

**b.** For that, it is necessary to invert the equations defining the spherical coordinates. Because of the multiple solutions for the angles, it is difficult for SymPy to solve these equations alone correctly. Ideally it would be enough to do:

```
eqn = [i-v for i,v in coords.items()]
inversecoords = sp.solve(eqn, rtp, dict=True)[0]
```

Unfortunately, according to the versions of SymPy, this gives either a solution which is correct only for some signs of $x$ and $y$ or no answer at all. The first step is to solve the last equation for $\theta$:

```
sol_t = sp.solve(eqn[2], t, dict=True)
```

then select the solution located in the interval $[0, \pi]$:

```
inversecoords = sol_t[1]
```

The second step is to substitute this solution into the first two equations and solve for $r$ and $\varphi$:

```
eqn12 = [sp.simplify(e.subs(inversecoords)) for e in
    eqn[0:2]]
sol_rphi = sp.solve(eqn12, (r,p), dict=True)
inversecoords.update(sol_rphi[0])
```

Note that the solution returned for $\varphi$ is the only solution, this last one is only correct for $x > 0$. To obtain the correct solution for the angle $\varphi$, the idea is to solve first for $\cos(\varphi)$ and $\sin(\varphi)$):

```
sol_cosp_sinp = sp.solve([e.subs({r:sol_rphi[0][r]}) for e in
    eqn12], (sp.cos(p),sp.sin(p)), dict=True)
```

then solve by hand with the atan2 function:

```
sol_p = sp.atan2(sp.cos(p),sp.sin(p)).subs(sol_cosp_sinp[0])
```

The last step is to create a simplification rule:

```
a,b,c = map(sp.Wild, "abc")
sol_p = sol_p.replace(sp.atan2(b/a,c/a), sp.atan2(b,c))
inversecoords[p] = sol_p
```

to finally obtain the inverse coordinates:

```
inversecoords = {k: v.subs(inversecoords) for k,v in
↳  inversecoords.items()}
inversecoords
```

The result should be:

$$\left\{\theta : \mathrm{acos}\left(\frac{z}{\sqrt{x^2+y^2+z^2}}\right),\; \varphi : \mathrm{atan}_2(x,y),\; r : \sqrt{x^2+y^2+z^2}\right\}.$$

Thus, the function allowing to convert into Cartesian coordinates an expression given in spherical coordinates is identical to the one allowing to do the opposite:

```
def to_cartesian(expr):
    return sp.simplify(expr.subs(inversecoords))
```

For example, to test it:

```
to_cartesian(r*sp.sin(t))
```

**c.** The first step is to compute the Jacobian matrix $\frac{\partial x_j}{\partial x_i'}$.

```
Jac = [[sp.diff(c,v) for c in coords.values()] for v in rtp]
sp.Matrix(Jac)
```

This allows us to define the scaling factors:

```
h = [sp.Matrix(line).norm() for line in Jac]
```

To simplify them, we have to say to SymPy that $\sin(\theta) \geq 0$:

```
assumption = sp.Q.positive(sp.sin(t))
h = [sp.refine(sp.simplify(i), assumption) for i in h]
```

Another way to do this is to take the roots of the elements of the diagonal of the metric:

```
metric = [[sum([sp.diff(c,v)*sp.diff(c,w) for c in
↳  coords.values()]) for v in rtp] for w in rtp]
metric = sp.simplify(sp.Matrix(metric))
```

**d.** According to the definition of the gradient in terms of the scaling factors:

```python
def gradient(expr):
    gradient = [sp.diff(expr,v)/h[i] for i,v in
    ↵    enumerate(rtp)]
    return sp.simplify(sp.Matrix(gradient))
```

To test this function:

```python
f = sp.Function("f")(r,t,p)
gradient(f)
```

**e.** For the Laplacian, we first compute the product of the factors $h_i$ in order to apply the definition:

```python
J = sp.prod(h)
def laplacian(expr):
    laplacian =
    ↵    1/J*sum([sp.diff(J/(h[i]**2)*sp.diff(expr,v),v) for
    ↵    i,v in enumerate(rtp)])
    return sp.expand(sp.simplify(laplacian))
```

To verify that the result is correct:

```python
f = sp.Function("f")(r,t,p)
laplacian(f)
```

**f.** We need to define a function depending only on the radial variable $r$ and then compute its Laplacian:

```python
G_radial = sp.Function("G")(r)
equation = laplacian(G_radial)
```

Finally the function dsolve allows to solve this ordinary differential equation:

```python
G_sol = sp.dsolve(equation, G_radial)
```

**g.** The first condition is easy to transform into spherical coordinates:

```python
cond1 = sp.limit(G_sol.rhs, r, sp.oo)
```

The second condition being seen in the sense of distributions, it is equivalent to:

$$\lim_{\varepsilon \to 0} \int_{B_\varepsilon} \Delta G(x)\, \mathrm{d}^3 x = 1,$$

where $B_\varepsilon$ denotes the $\mathbb{R}^3$ ball of radius $\varepsilon$ and centered at the origin. By integrating the left side by parts:

$$\lim_{\varepsilon \to 0} \int_{\partial B_\varepsilon} \nabla G(x) \cdot n\, \mathrm{d}^3 x = 1,$$

and finally by expressing this last integral in spherical coordinates:

$$\lim_{\varepsilon \to 0} \int_0^\pi \int_0^{2\pi} \left(\boldsymbol{\nabla} G(\varepsilon) \cdot \boldsymbol{e}_r\right) J(\varepsilon, \theta, \varphi) \, d\varphi d\theta = 1 \,.$$

Since $\varepsilon$ can be replaced by $r$ in this last expression, we define the normal to the surface in spherical coordinates, then we compute the integral in spherical coordinates, and finally the limit:

```
normal = sp.Matrix([1,0,0])
integral = sp.integrate(normal.dot(gradient(G_sol.rhs))*J,
 ↳ (t,0,sp.pi), (p,0,2*sp.pi))
cond2 = sp.limit(integral, r, 0) - 1
```

**h.** It is sufficient to solve the equations given by the two previous conditions:

```
constants = sp.solve([cond1, cond2], [sp.Symbol('C1'),
 ↳ sp.Symbol('C2')])
```

then substitute in the solution:

```
G_spherical = G_sol.rhs.subs(constants)
```

To obtain the Green's function in Cartesian coordinates, we use the function defined previously:

```
G = to_cartesian(G_spherical)
```

**i.** First, we define a new set of Cartesian coordinates $x_0$ and of spherical coordinates $(r_0, \theta_0, \varphi_0)$:

```
x0, y0, z0 = sp.symbols("x_0 y_0 z_0", real=True)
xyz0 = [x0,y0,z0]
r0, t0, p0 = sp.symbols(r"r_0 \theta_0 \varphi_0", real=True,
 ↳ positive=True)
rtp0 = [r0,t0,p0]
coords0 = {i.subs({x:x0, y:y0, z:z0}): v.subs({r:r0, t:t0,
 ↳ p:p0}) for i,v in coords.items()}
inversecoords0 = {i.subs({r:r0, t:t0, p:p0}): v.subs({x:x0,
 ↳ y:y0, z:z0}) for i,v in inversecoords.items()}
```

Then, we define the Green's function $G(\boldsymbol{x} - \boldsymbol{x}_0)$ and convert it into spherical coordinates:

```
green_cartesian = G.subs({x: x-x0, y:y-y0, z:z-z0})
green_spherical =
 ↳ sp.simplify(green_cartesian.subs(coords).subs(coords0))
```

To do the asymptotic development:

```
serie = sp.series(green_spherical, r, sp.oo, 3)
```

Then, we define a function g depending only on $r_0$ and $\theta_0$ and the factor $J(r_0, \theta_0, \varphi_0)$:

```
g = sp.Function("g")(r0,t0)
J0 = J.subs({r:r0, t:t0, p:p0})
```

To calculate the asymptotic first order decreasing as $r^{-1}$:

```
int1 = serie.coeff(r,-1)
asy1 = sp.integrate(int1*g*J0, (p0,0,2*sp.pi))
sp.expand(asy1)
```

and identically for the second order decreasing as $r^{-2}$:

```
int2 = serie.coeff(r,-2)
asy2 = sp.integrate(int2*g*J0, (p0,0,2*sp.pi))
sp.trigsimp(sp.expand(asy2), method='combined')
```

Thus, the final result can be written mathematically as:

$$f(r, \theta, \varphi) = \frac{M_0}{r} + \frac{M_1 \cos(\theta)}{r^2} + O\left(\frac{1}{r^3}\right),$$

where $M_0$ and $M_1$ are the real numbers given by:

$$M_0 = \frac{-1}{2} \int_0^\infty \int_0^\pi g(r, \theta) r^2 \sin(\theta)\, d\theta dr,$$

$$M_1 = \frac{-1}{4} \int_0^\infty \int_0^\pi g(r, \theta) r^3 \sin(2\theta)\, d\theta dr.$$

These last two real numbers can also be calculated in Cartesian coordinates:

```
car = [sp.simplify(asy/J0/g/(4*sp.pi)) for asy in [asy1,
⤷   asy2]]
[x.subs(inversecoords0).subs(inversecoords) for x in car]
```

which gives mathematically:

$$M_0 = \frac{-1}{8\pi} \int_{\mathbb{R}^3} g(x)\, d^3x, \qquad M_1 = \frac{-1}{8\pi} \int_{\mathbb{R}^3} x_3\, g(x)\, d^3x.$$

# CHAPTER 9

# Root Finding

The aim of this series of exercises is to determine the roots of a function in an approximate way, in particular by Newton's method. This allows in particular to find approximate solutions of nonlinear equations. This method is fundamental both from a numerical and an analytical point of view.

### Concepts covered

- Newton's method in one and several dimensions

- Jacobian matrix

- Newton's method attractor

- fractal set

- optimization by parallelization

- nonlinear differential equation

- finite differences

DOI: 10.1201/9781003565451-9

# EXERCISES

## EXERCISE 9.1   NEWTON'S METHOD IN ONE DIMENSION

In one dimension, Newton's method consists in finding an approximate solution of a single equation. This equation can be put in the general form $F(x) = 0$, where $F : \mathbb{R} \to \mathbb{R}$ is a fairly regular function; so, the goal is to find a zero of the function $F$. The equation $F(x) = 0$ is equivalent (if $F'(x) \neq 0$) to the equation $G(x) = x$, where $G$ is the function defined by:

$$G(x) = x - \frac{F(x)}{F'(x)}.$$

Newton's method consists in finding a fixed point of $G$, i.e., solving $G(x) = x$ by successive iterations:

$$x_{i+1} = G(x_i) = x_i - \frac{F(x_i)}{F'(x_i)},$$

from an initial value $x_0 \in \mathbb{R}$. When the sequence $(x_i)_{i \in \mathbb{N}}$ converges, then the limit $x$ is a solution of $G(x) = x$ therefore of $F(x) = 0$.

**a.** Write a function `newton1d(F, DF, x0, eps=1e-10, N=1000)` that given a function $F$, its derivative $F'$, and an initial value $x_0$ computes Newton's iterations until $|F(x_i)| < \varepsilon$ and returns $x_i$. If $N$ iterations were not enough to reach this convergence criterion, then return an error.

**b.** Using the function defined above, find an approximate solution of the equation $e^{-x} = x$.
*Answer: The solution is approximately given by $x = 0.56714$.*

**c.** Without using the function `sqrt`, `log`, or fractional powers, define a function `root(x,n)` that computes $\sqrt[n]{x}$.
Sometimes, the derivative of the function $F$ cannot be calculated analytically, so it is necessary to approximate it numerically:

$$F'(x_i) \approx \frac{F(x_i) - F(x_{i-1})}{x_i - x_{i-1}},$$

which leads to the secant method:

$$x_{i+1} = x_i - F(x_i)\frac{x_i - x_{i-1}}{F(x_i) - F(x_{i-1})} = \frac{x_{i-1}F(x_i) - x_i F(x_{i-1})}{F(x_i) - F(x_{i-1})},$$

where $x_0$ and $x_1$ must be chosen.

**d.** Write a method `secant1d(F, x0, x1, eps=1e-10, N=1000)` implementing the secant method and test it on the previous example.

## EXERCISE 9.2   NEWTON'S METHOD IN SEVERAL DIMENSIONS

Newton's method in one dimension is easily generalized to several dimensions to solve equations of the form $F(x) = 0$, where $F : \mathbb{R}^n \to \mathbb{R}^n$ is a fairly regular function. Conceptually, the method is identical: the equation $F(x) = 0$ is equivalent to $G(x) = x$ with the function $G$ defined by:

$$G(x) = x - \left(F'(x)\right)^{-1} F(x),$$

where $F'(x)$ denotes the Jacobian matrix of size $n \times n$ of $F$ in $x$. Thus, Newton's iterations are written:

$$x_{i+1} = x_i - \left(F'(x_i)\right)^{-1} F(x_i),$$

**a.** Write a function newton(F, DF, x0, eps=1e-12, N=10000) implementing Newton's method in more than one dimension.

*Hint: To have an optimal performance, one should not invert the Jacobian matrix but solve a linear system with $F(x_i)$ as second member.*

**b.** Use the previous function to solve the following system:

$$\cos(x) = \sin(y), \qquad\qquad e^{-x} = \cos(y).$$

*Answer: A solution is approximately given by $x = 0.58853$ and $y = 0.98226$.*


## EXERCISE 9.3   NEWTON'S METHOD ATTRACTOR

The goal of this exercise is to solve the equation $z^3 = 1$ in the complex plane using Newton's method and to analyze to which of the three roots of unity the method will converge depending on the choice of the initial point $z_0$.

**a.** If necessary, adapt the function newton1d so that it also applies to complex numbers and test it to solve $z^3 = 1$ from different values of $z_0$.

**b.** Determine for each $z_0 \in \{x_0 + iy_0 : x_0 \in [-3, 3]$ and $y_0 \in [-3, 3]\}$ to which root of unity Newton's method will converge. Represent graphically this set as in Figure 9.1.

*Hint: NumPy's meshgrid function can be useful to construct the matrix corresponding to the set of $z_0$.*

**c.** ! The previous method has the disadvantage of proceeding sequentially to the calculation for each value of $z_0$, which makes this evaluation rather slow. Propose a new implementation allowing to compute in parallel all the values of $z_0$ using NumPy indexing.

*Hint: To speed-up the method even more, Newton's iterations of $F(z) = z^3 - 1$ can be calculated by hand:*

$$z_{n+1} = \frac{1}{3z_n^2} + \frac{2z_n}{3}.$$

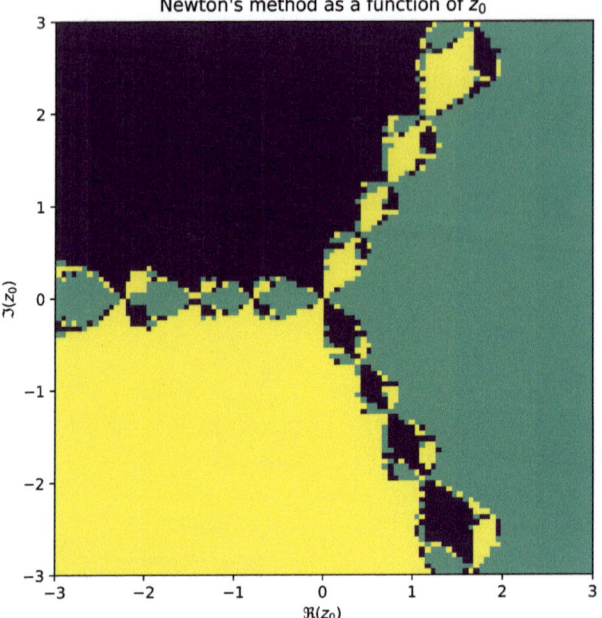

Figure 9.1 Graphical representation of the convergence of Newton's method as a function of the initial value $z_0$.

## EXERCISE 9.4 NONLINEAR DIFFERENTIAL EQUATION (!!)

The goal is to solve the following differential equation with boundary conditions:

$$u''(x) + u^3(x) = \sin(x), \qquad\qquad u(0) = u(2\pi) = 0,$$

on the interval $[0, 2\pi]$. This equation is a simplified model for a nonlinear Schrödinger equation.

The method used is finite differences that consists in looking for the values of $u$ at the points $x_n = \frac{2\pi n}{N}$ for $n = 0, 1, \dots, N$. The unknowns are then the numbers $u_n = u(x_n)$ and form a vector of dimension $N + 1$. The finite difference method consists in approximating the second derivative by:

$$u''(x) \approx \frac{u(x + h) - 2u(x) + u(x - h)}{h^2},$$

when $h$ is small. Taking $h = \frac{2\pi}{N}$, then:

$$u''(x_n) \approx \frac{u_{n+1} - 2u_n + u_{n-1}}{h^2},$$

and so the initial equation is approximated by:

$$\frac{u_{n+1} - 2u_n + u_{n-1}}{h^2} + u_n^3 = \sin(x_n), \qquad u_0 = u_N = 0,$$

for $n = 1, 2, \dots, N-1$. This equation can be seen as an equation of the type $F(\boldsymbol{u}) = \boldsymbol{0}$ for $\boldsymbol{u} = (u_n)_{n=0}^{N+1}$ and thus be solved by Newton's method.

**a.** Show the following approximation:

$$u''(x) = \frac{u(x+h) - 2u(x) + u(x-h)}{h^2} + O(h^2) \quad \text{as} \quad h \to 0.$$

*Hint: Use Taylor's theorem.*

**b.** Define a vector x representing $N + 1$ evenly spaced points in $[0, 2\pi]$ and h the distance between the points, with, for example, $N = 200$.

**c.** Define a function F(u) representing the function $F : \mathbb{R}^{N+1} \to \mathbb{R}^{N+1}$ allowing to put the approximated equation in the form $F(\boldsymbol{u}) = \boldsymbol{0}$.
*Hint: To have a fast implementation, it is imperative to use the NumPy slicing instead of a loop to build F.*

**d.** Define a function DF(u) representing the Jacobian of the previous function.
*Hint: The Jacobian is the derivative of $F(\boldsymbol{u}) = F(u_0, u_1, \dots, u_N)$ with respect to $\boldsymbol{u} = (u_0, u_1, \dots, u_N)$, i.e.:*

$$F'(\boldsymbol{u}) = \Big(\partial_0 F(\boldsymbol{u}) \quad \partial_1 F(\boldsymbol{u}) \quad \partial_2 F(\boldsymbol{u}) \quad \cdots \quad \partial_{N-1} F(\boldsymbol{u}) \quad \partial_N F(\boldsymbol{u})\Big),$$

*and can be calculated explicitly by hand.*

**e.** Use the newton function defined earlier to calculate an approximate solution of the equation. By changing the initial values, is it possible to find other solutions?
*Hint: Try with the initial data $u_0(x) = (1 + k)\sin(kx)$ for $k = 1, 2, 3, 4$ as the starting point of Newton's method.*

# SOLUTIONS

---

## SOLUTION 9.1   NEWTON'S METHOD IN ONE DIMENSION

**a.** In order not to create an infinite `while` loop if Newton's method does not converge, the simplest way is to use a `for` loop even if you have to exit it when convergence is reached:

```python
def newton1d(F, DF, x0, eps=1e-12, N=10000):
    x = x0
    for i in range(N):
        # calculate F(x) and DF(x)
        Fx = F(x)
        DFx = DF(x)
        # test if the precision is sufficient
        if abs(Fx) < eps:
            return x
        # test that the derivative is not too small
        if abs(DFx) < eps:
            raise Exception(f"Derivative DF = {DFx} too
              small")
        # otherwise Newton's iteration
        x -= Fx/DFx
    # if the loop ends, one has not converged (yet)
    raise Exception(f"The error after {N} iterations is
      {abs(Fx)} > {eps}")
```

**b.** The equation $e^{-x} = x$ can be put in the form $F(x) = 0$ with $F(x) = e^{-x} - x$:

```python
import math
F = lambda x: math.exp(-x)-x
DF = lambda x: -math.exp(-x)-1
newton1d(F, DF, 1)
```

**c.** Just solve the equation $t^n = x$ with Newton's method:

```python
def racine(x,n):
    F = lambda t: t**n - x
    DF = lambda t: n*t**(n-1)
    return newton1d(F, DF, x)
```

To test:

```python
[racine(17,i) for i in range(1,17)]
```

**d.** The assignment of a tuple allows to get rid of a temporary variable:

```python
def secant1d(F, x0, x1, eps=1e-12, N=10000):
    x_ = x0 # old x
    x = x1  # new x
    for i in range(N):
        # calculate F(x)
        Fx = F(x)
        Fx_ = F(x_)
        # test if the precision is sufficient
        if abs(Fx) < eps:
            return x
        # test that the derivative is not too small
        if abs(Fx-Fx_) < eps:
            raise Exception(f"Approximate derivative {Fx-Fx_}
                too small")
        # otherwise secant iteration
        x, x_ = (x_*Fx - x*Fx_)/(Fx - Fx_), x
    # if the loop ends, one has not converged (yet)
    raise Exception(f"The error after {N} iterations is
        {abs(Fx)} > {eps}")
secant1d(F, 0, 1)
```

## SOLUTION 9.2    NEWTON'S METHOD IN SEVERAL DIMENSIONS

**a.** The choice is made to treat $x$ as a NumPy vector, which allows the use of the NumPy linear system solver:

```python
import numpy as np
def newton(F, DF, x0, eps=1e-12, N=10000):
    x = x0.copy()
    for i in range(N):
        # calculate F(x) and DF(x)
        Fx = F(x)
        DFx = DF(x)
        # test if the precision is sufficient
        if np.linalg.norm(Fx) < eps:
            return x
        # test that the derivative is not too small
        if np.linalg.norm(DFx) < eps:
            raise Exception(f"Derivative  |DF| =
                {np.linalg.norm(DFx)} too small")
        # solve d = DFx^{-1} Fx then Newton's iteration
        x -= np.linalg.solve(DFx, Fx)
    # if the loop ends, one has not converged (yet)
    raise Exception(f"The error after {N} iterations is
        {np.linalg.norm(Fx)} > {eps}")
```

**b.** The previous equation can be written as $F(\boldsymbol{x}) = \boldsymbol{0}$ with:

$$F(x, y) = \begin{pmatrix} \cos x - \sin y \\ e^{-x} - \cos y \end{pmatrix},$$

Thus:

$$F'(x, y) = \begin{pmatrix} -\sin x & -\cos y \\ -e^{-x} & \sin y \end{pmatrix},$$

and therefore:

```
F = lambda x: np.array([np.cos(x[0]) - np.sin(x[1]),
    np.exp(-x[0]) - np.cos(x[1])])
DF = lambda x: np.array([[-np.sin(x[0]), -np.cos(x[1])],
    [-np.exp(-x[0]), np.sin(x[1])]])
newton(F, DF, np.array([0.,0.]))
```

## SOLUTION 9.3   NEWTON'S METHOD ATTRACTOR

**a.** The previously defined function does not need to be modified:

```
F = lambda z: z**3-1
DF = lambda z: 3*z**2
[newton1d(F, DF, z0, eps=1e-15) for z0 in
    [1,1+2j,-3-1j,-3+1j]]
```

**b.** First we define the set of $z_0$:

```
nb = 100
lst = np.linspace(-3,3,nb)
x0, y0 = np.meshgrid(lst,lst)
z0 = x0 + 1j*y0
```

Then, we build the set of results of Newton's iterations:

```
out = z0.copy()
for i in range(nb):
    for j in range(nb):
        out[i,j] = newton1d(F, DF, z0[i,j])
```

Finally, using Matplotlib, we represent the argument of each complex number to obtain Figure 9.1:

```
import matplotlib.pyplot as plt
plt.figure(figsize=(6,6))
plt.title("Newton's method as a function of $z_0$")
plt.xlabel(r"$\Re(z_0)$")
```

```
plt.ylabel(r"$\Im(z_0)$")
plt.imshow(np.angle(out), extent=[-3,3,-3,3])
```

The conclusion is that in a neighborhood of each of the three roots, Newton's method converges well toward this root; on the contrary, as soon as one moves too far away from one of the roots, then Newton's method can converge toward one or the other of the roots in a fractal way.

**c.** An array of the same size as $z_0$ is iterated successively. To gain even more speed, only the entries of the array that have not yet converged are iterated:

```
def parallel(z0, eps=1e-12, N=1000):
    z = z0.copy()
    # boolean array of the z that have not diverged
    cond = np.abs(z**3-1) > eps
    # iterations
    for i in range(N):
        # select the z that have not converged
        cond[cond] = np.abs(z[cond]**3-1) > eps
        # terminate when all z have converged
        if cond.any() == False:
            return z
        # Newton's iteration
        z[cond] = 1/(3*z[cond]**2) + 2*z[cond]/3
    # some z have not converged
    raise Exception("Some z have not converged")
```

For $1\,000 \times 1\,000$ values of $z_0$, the method computing the iterates for each $z_0$ separately and successively takes about 10 seconds to complete. The current method takes about 0.4 seconds to compute the result:

```
out = parallel(z0)
```

This can be used to generate a high-resolution fractal:

```
# calculate the points
nb = 4000
lst = np.linspace(-3,3,nb)
x0, y0 = np.meshgrid(lst,lst)
z0 = x0 + 1j*y0
out = parallel(z0)
# create the figure with 1 pixel per calculated point
fig = plt.figure(figsize=(nb/100,nb/100))
ax = fig.add_axes([0, 0, 1, 1])
ax.axis('off')
ax.imshow(np.angle(out))
ax.set(xlim=[0, nb], ylim=[nb, 0], aspect=1)
plt.savefig('newton.png', dpi=100, transparent=True)
plt.clf()
```

## SOLUTION 9.4   NONLINEAR DIFFERENTIAL EQUATION (!!)

**a.** With the Taylor's theorem,

$$u(x + h) = u(x) + u'(x)h + \frac{1}{2}u''(x)h^2 + \frac{1}{6}u'''(x)h^3 + O(h^4),$$

$$u(x - h) = u(x) - u'(x)h + \frac{1}{2}u''(x)h^2 - \frac{1}{6}u'''(x)h^3 + O(h^4),$$

and therefore:

$$u(x + h) - 2u(x) + u(x - h) = u''(x)h^2 + O(h^4).$$

**b.** This is the perfect example of how to use the `linspace` function of NumPy:

```
N = 200
x = np.linspace(0,2*np.pi,N+1)
h = x[1]-x[0]
```

**c.** Mathematically the function $F$ is defined by:

$$F(\boldsymbol{u}) = \begin{pmatrix} u_0 \\ \dfrac{u_0 - 2u_1 + u_2}{h^2} + u_1^3 - \sin(x_1) \\ \dfrac{u_1 - 2u_2 + u_3}{h^2} + u_2^3 - \sin(x_2) \\ \vdots \\ \dfrac{u_{N-2} - 2u_{N-1} + u_N}{h^2} + u_{N-1}^3 - \sin(x_{N-1}) \\ u_N \end{pmatrix}$$

and is therefore implemented in the following way:

```
def F(u):
    # result
    out = np.zeros_like(u)
    # second order finite difference
    upp = (u[:-2] - 2*u[1:-1] + u[2:])/h**2
    # define F inside the interval
    out[1:-1] = upp + u[1:-1]**3 - np.sin(x[1:-1])
    # force zero on the boundaries of the interval (this
    ↳   should always be the case)
    out[0] = u[0]
    out[-1] = u[-1]
    return out
```

**d.** Mathematically the Jacobian is given by:

$$F'(\boldsymbol{u}) = \begin{pmatrix} 1 & 0 & 0 & 0 & 0 & 0 \\ \dfrac{1}{h^2} & -\dfrac{2}{h^2} + 3u_1^2 & \dfrac{1}{h^2} & 0 & 0 & 0 \\ 0 & \dfrac{1}{h^2} & -\dfrac{2}{h^2} + 3u_2^2 & \ddots & 0 & 0 \\ 0 & 0 & \ddots & \ddots & \dfrac{1}{h^2} & 0 \\ 0 & 0 & 0 & \dfrac{1}{h^2} & -\dfrac{2}{h^2} + 3u_{N-1}^2 & \dfrac{1}{h^2} \\ 0 & 0 & 0 & 0 & 0 & 1 \end{pmatrix}$$

and therefore:

```python
def DF(u):
    # matrix of the second order derivative
    A = -2*np.eye(N+1) + np.eye(N+1,k=1) + np.eye(N+1,k=-1)
    A /= h**2
    # add the nonlinear parts on the diagonal
    A += np.diag(3*u**2)
    # corrections due to boundary conditions
    A[0,0] = 1 ; A[0,1] = 0 ; A[-1,-1] = 1 ; A[-1,-2] = 0
    return A
```

Note that this implementation is not optimal when $N$ is very large. Indeed, the Jacobian matrix is tridiagonal and so it is useless to store the whole matrix; storing the three diagonals is enough. This can be done using the sparse matrices defined in the sparse module of SciPy.

**e.** Starting with the initial data $u_0(x) = 0$ for Newton's method, we find a solution:

```python
u0 = np.zeros(N+1)
sol = newton(F, DF, u0, eps=1e-8)
plt.plot(x,sol)
```

By changing the initial data to $u_0(x) = (1 + k)\sin(kx)$ for $k = 1, 2, 3, 4$, then we get other solutions as shown in Figure 9.2:

```python
plt.figure(figsize=(8,5))
plt.title(r"Solutions of the ODE $u^{\prime\prime} + u^3 =
    \sin(x)$")
for k in range(1,5):
    u0 = (1+k)*np.sin(k*x)
    sol = newton(F, DF, u0, eps=1e-8)
    plt.plot(x, sol, label = f"{k}")
    plt.xticks([0,np.pi/2, np.pi, 3*np.pi/2,2*np.pi],
        [r'$0$',r'$\frac{\pi}{2}$', r'$\pi$',
        r'$\frac{3\pi}{2}$', r'$2\pi$'])
```

```
    plt.xlim([0,2*np.pi])
plt.legend()
```

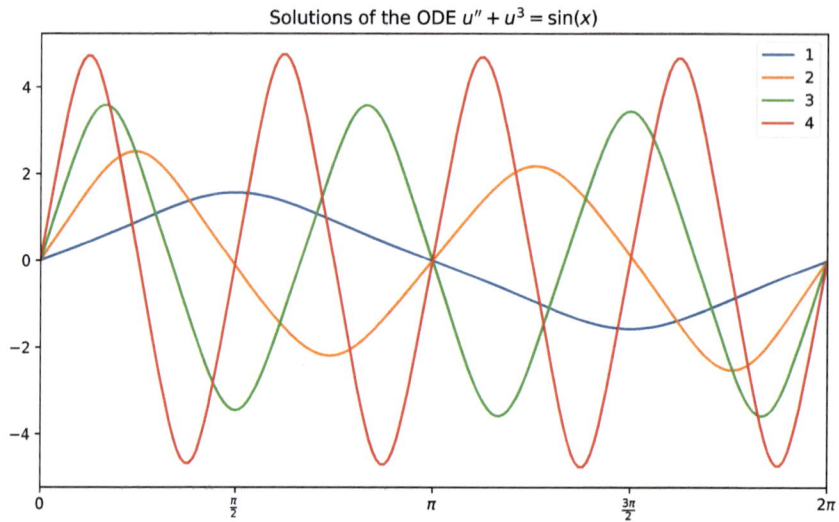

Figure 9.2   First solutions of the nonlinear differential equation $u'' + u^3 = 0$.
There seems to be a discrete infinity of solutions.

# CHAPTER 10

# Probability and Statistics

In a first step, the statistics of the proportion of numbers beginning with a certain digit will be studied. Then in a second step, important probabilistic models will be introduced and simulated, such as random walks, illustrations of the central limit theorem, or percolation.

## Concepts covered

- statistics and probability

- random harmonic series

- random walk

- central limit theorem

- random vectors

- percolation

- phase transition

- histograms

- optimization by compilation

DOI: 10.1201/9781003565451-10

# EXERCISES

## EXERCISE 10.1   HARMONIC SERIES OF RANDOM SIGN

The goal of this exercise is to simulate the convergence of a harmonic series whose sign is drawn randomly. More precisely, if $(X_i)_{i \in \mathbb{N}}$ is a sequence of independent random variables worth $-1$ or $1$ with probability $\frac{1}{2}$, then we define the partial sum:

$$W_0 = 0, \qquad W_n = \sum_{i=1}^{n} \frac{X_i}{i},$$

and the question is to determine if the sequence $(W_n)_{n \in \mathbb{N}}$ converges and if so toward what.

**a.** Write a function `sign()` that simulates the random variable $X_i$.

**b.** Write a function `simulate(n)` that returns a realization of $(W_0, W_1, \dots, W_n)$.

**c.** Plot the function $n \mapsto W_n$ for different realizations, for example, for $0 \le n \le 1\,000$ and make a conjecture about the convergence of the sequence $(W_n)_{n \in \mathbb{N}}$.

**d.** ! Determine the histogram of $W_{1\,000}$ for $10^4$ or $10^5$ realizations to get an idea of the law of the limiting random variable.

## EXERCISE 10.2   GAMBLER'S RUIN

The goal is to simulate the evolution of the amount of money of a gambler playing heads or tails. At each toss, the player wins one euro if it is heads and loses one if it is tails. The probability of getting tails is noted $p$, that of getting heads $q$. In particular, $p = q = \frac{1}{2}$ if the coin is balanced.

Mathematically, the sum $S_i$ owned by the player at time $i$ is given by a random walk:

$$S_i = \begin{cases} 0, & \text{if } S_{i-1} = 0, \\ S_{i-1} + X_i, & \text{if } S_{i-1} \ge 1, \end{cases}$$

where $(X_i)_i$ are independent random variables of law $\mathbb{P}(X_i = 1) = p$ and $\mathbb{P}(X_i = -1) = q$.

**a.** Write a function `simulate(p,k,N)` that generates a realization of length $N$ of the process from $S_0 = k$, i.e., returns $(S_0, S_1, S_2, \dots, S_N)$. Represent graphically several realizations.

**b.** Simulate a player who, starting with a sum $k$, plays until he loses everything or has the amount $n \ge k$.

c. If $T$ is the time at which the game stops, *i.e.*, when $S_T = 0$ or $S_T = n$, recover by simulation the theoretical results on the average time:

$$\mathbb{E}(T) = \begin{cases} k(n-k), & \text{if } p = q, \\ \dfrac{n}{p-q}\dfrac{1-\rho^k}{1-\rho^n} - \dfrac{k}{p-q}, & \text{if } p \neq q, \end{cases}$$

and the place of exit:

$$\mathbb{P}(S_T = 0) = \begin{cases} \dfrac{n-k}{n}, & \text{if } p = q, \\ \dfrac{\rho^k - \rho^n}{1-\rho^n}, & \text{if } p \neq q, \end{cases}$$

where $\rho = q/p$. For this, we can plot these quantities as a function of $p$ or just consider the case $p = q = \dfrac{1}{2}$.

## EXERCISE 10.3   PÓLYA URN

An urn initially contains (at $t = 0$) $r_0$ red balls and $b_0$ white balls. At each time, we pick a ball uniformly at random from the urn. This ball is then returned to the urn and a ball of the same color is added. Such a system is called a *Pólya urn*. The purpose of this exercise is to study the behavior of the fraction of red balls in the urn, *i.e.*, the number of red balls out of the total number. We will call $r_n$ and $b_n$, respectively, the number of red and white balls in the urn at time $n$.

a. Write a function density taking as argument a tuple representing the number of red and white balls in an urn, and which returns the density of red balls.

We want to recursively construct the distribution of the number of red balls at time $n$, *i.e.*, the list of probabilities that the number of red balls is equal to a given integer $k$ (which will be the index of the list). This is done by writing two functions: next_dist_red, which takes as argument the distribution at time $n$ and returns the one at time $n + 1$, which is thus the function that does all the work, and dist_red, which is the wrapper function, taking as argument $r_0$, $b_0$, and time $n$ and returning the distribution at time $n$ by a recursive call. We will use the following useful facts (make a small drawing):

- The distribution passed as an argument to next_dist_red is a list r, and r[k] represents the probability of having k red balls in the urn at time $n$. The indices for r vary from 0 to the total number s of balls at time $n$.

- At time $n + 1$, to have k red balls, we need:

    - either having had $k$ red balls at the previous time and not having drawn a red ball;

    – or having had $k - 1$ red balls at the previous time and having drawn a red ball.

- If $n = 0$, the result of `dist_red` is completely deterministic and the coefficients of the list are only 0 and 1, depending on $r_0$ and $b_0$.

**b.** Write the functions `next_dist_red` and `dist_red` using the directions provided. Look at the result of `dist_red(0,1,n)` and `dist_red(1,1,n)` for different values of n (1, 2, 5, 10, 20, ... ) and comment.

Rather than theoretically computing for each $n$ the sequence of theoretical probabilities, we will do statistics on a large number of Pólya urn realizations after a large number of steps. For this, we need a function to evolve a Pólya urn.

**c.** Define a function `polya_step(r,b)` which, given the composition of an urn passed as two parameters r and b, returns the (random) evolution after one step of the composition of the urn as a tuple. Also define a function `polya(r0,b0,N)` taking as arguments $r_0, b_0$, and $N$ as parameters and returning the (random) composition of a Pólya urn after $N$ steps, also as a tuple.

**d.** Write a function `data_rdens_polya(r0,b0,N,nbexp)` that returns a list of length nbexp containing the densities of nbexp realizations of Pólya urns at time N initialized with r0 red balls and b0 white balls.

**e.** Store the result of `data_rdens_polya(2,3,1000,10_000)` in a variable and draw a histogram to see the distribution of densities. Be careful, we want the heights of the bars to be normalized so that their surface represents the proportion of points, and not so that they give the number of points per bin.
*Hint: A good rule of thumb is to choose the number of bins for a histogram of the order of the square root of the number of points. See the documentation of the `hist` function of Matplotlib.*

## EXERCISE 10.4   CENTRAL LIMIT THEOREM

The central limit theorem establishes the convergence of the sum of a sequence of random variables toward the normal distribution. Intuitively, this result states that a sum of identical and independent random variables tends (under certain conditions) toward a Gaussian random variable. Here's how it works:

**Theorem:** Let $(X_n)$ be a sequence of independent real random variables of the same distribution with expectation $\mu$ and standard deviation $\sigma \neq 0$. Let $(\bar{X}_n)$ be the sequence defined by:

$$\bar{X}_n = \frac{1}{n} \sum_{k=1}^{n} X_k.$$

For $n$ large enough, the distribution of $\bar{X}_n$ can be approximated by the normal distribution $\mathcal{N}(\mu, \frac{\sigma^2}{n})$.

The aim of this exercise is to check whether this theorem is valid for different laws of probability:

- **Poisson distribution**: discrete distribution on $\mathbb{N}$, of parameter $\lambda$, defined by:

$$\mathbb{P}(X = k) = \exp(-\lambda)\frac{\lambda^k}{k!}, \quad \forall k \in \mathbb{N}.$$

- **Normal distribution**: continuous distribution on $\mathbb{R}$, of parameters $m$ and $\sigma$, defined by density:

$$\frac{1}{\sqrt{2\pi\sigma^2}} \exp\left(-\frac{(x-m)^2}{2\sigma^2}\right), \quad \forall x \in \mathbb{R}.$$

- **Cauchy distribution**: continuous distribution on $\mathbb{R}$, of parameters $a$ and $\gamma$, defined by density:

$$\frac{1}{\pi\gamma\left(1 + \left(\frac{x-a}{\gamma}\right)^2\right)}, \quad \forall x \in \mathbb{R}.$$

**a.** Define a function `normal_density(x, mu var)` taking as an argument :

- `x` (array): an array of floating-point numbers;

- `mu` (floating number): average $\mu$;

- `var` (strictly positive floating number): variance $\sigma^2$;

and which returns the density of the normal distribution evaluated for each number $x$ in `x`:

$$\mathcal{N}(\mu, \sigma^2) = \frac{1}{\sqrt{2\pi\sigma^2}} \exp\left(-\frac{(x-\mu)^2}{2\sigma^2}\right).$$

**b.** Look at the documentation for the `numpy.random.poisson` function and generate 10 random values according to Poisson distribution with parameter $\lambda = 2$.

**c.** Write a function `samples_poisson(lam, N, M)` which takes as arguments:

- `lam` (strictly positive real): parameter $\lambda$ for the Poisson distribution;

- `N` (strictly positive integer): number of experiments;

- `M` (strictly positive integer): number of random variables generated for each experiment;

and which generates `N` experiments with `M` random variables generated per experiment, and returns:

- the average value (floating number) over the `N * M` random variables generated;

- the standard deviation (floating number) on the N ＊ M random variables generated;

- a numpy vector of size N where each element is the mean of the random variables in an experiment.

For the values lam=2 and N=10_000, run samples_poisson(lam, N, M) for $M \in \{10, 100, 1\,000\}$ and save the results in variables.

**d.** For each value of M, display the distribution of the numpy vector containing the means of each experiment, as well as the distribution of the expected normal distribution if the central limit theorem is verified. You can use Matplotlib hist function to display the histogram of a table, with the bins=50 parameter to set the number of columns and density=True to display a probability distribution. Use the empirical means and standard deviations returned by the samples_poisson function for normal distribution parameters. Choose relevant values for the x-axis boundary values. Make a hypothesis on the validity of the central limit theorem for the Poisson distribution.

**e.** Repeat questions **c.** and **d.** for the normal distribution. Using the function numpy.random.normal, generate averages for loc=2, scale=1, N=10_000, and $M \in \{10, 100, 1\,000\}$. Plot the histograms and make a hypothesis about the validity of the central limit theorem for the normal distribution.

**f.** Repeat questions **c.** and **d.** for the Cauchy distribution with $a = 0$ and $\gamma = 1$. To do this, use the function numpy.random.standard_cauchy. Use bins=np.arange(-10, 10.1, 0.1) and make an assumption about the validity of the central limit theorem for the Cauchy distribution.

## EXERCISE 10.5  RANDOM GENERATION OF UNIT VECTORS

The aim of this exercise is to find an efficient method for randomly generating unit vectors in $\mathbb{R}^n$ according to a uniform distribution. We will start with the case $n = 2$, where a real vector can be represented by a complex number.

**a.** Consider the following strategy for randomly generating a unit vector in $\mathbb{R}^2$:

1. Generate $x$ randomly according to the uniform distribution on $[-1, 1]$;

2. Generate $y$ randomly according to the uniform distribution on $[-1, 1]$;

3. Return the unit complex $z = \dfrac{x}{\sqrt{x^2+y^2}} + \dfrac{y}{\sqrt{x^2+y^2}}i$.

Write a function generate_complex which takes as argument a positive integer $N$ and returns a NumPy array of size $N$, where each element is a complex generated by the above strategy.
*Hint: It is much more efficient to generate $N$ random variables directly using the* size *argument of the function* numpy.random.uniform *than to generate them one by one with a* for *loop.*

**b.** For $n = 2$, one way to check whether the distribution of vectors is uniform is to look at the distribution of angles/arguments of the complex numbers (*i.e.*, $\arg z$): this should be uniform. Use the `generate_complex` function with $N = 10^6$ and display the distribution of angles/arguments. Does the strategy described in the previous question generate unit vectors uniformly?
*Hint: You can use the function* `numpy.angle`.

**c.** The following modification to the previous strategy is proposed:

1. Generate $x$ randomly according to the uniform distribution on $[-1, 1]$;

2. Generate $y$ randomly according to the uniform distribution on $[-1, 1]$;

3. If $x^2 + y^2 \leq 1$, return the unit complex $z = \dfrac{x}{\sqrt{x^2+y^2}} + \dfrac{y}{\sqrt{x^2+y^2}}i$ (otherwise return nothing at all).

Write a function `generate_complex_monte_carlo` which takes as argument a positive integer $N$ corresponding to the number of candidate vectors and returns a NumPy array of size $n \leq N$, where each element is a complex generated by the above strategy. **Warning:** $n$ is random and therefore cannot be determined in advance.
Repeat the previous question and display the angle/argument distribution of the complexes generated by this strategy. Does this strategy generate unit vectors uniformly?

**d.** We now ask how many candidate complexes must be generated on average to accept $n$ of them, which is equivalent to determining how many complexes are accepted on average from the $N$ candidates. The law of large numbers means that the ratio converges to the probability that a candidate vector will be accepted. This probability is equal to the ratio of the inclusion area $\pi$ (the unit circle) to the total area 4 (the square $[-1, 1]^2$), *i.e.*, $\dfrac{\pi}{4}$. Compare the ratio $\dfrac{n}{N}$ to $\dfrac{\pi}{4}$.

**e.** We now consider the general case $n \geq 2$. The strategy considered is the same as in question **c.**:

1. Generate a vector $x = (x_1, \dots, x_n)$, where each $x_k$ is randomly and independently generated according to the uniform distribution on $[-1, 1]$;

2. If $\|x\|_2 \leq 1$, return the unit vector $\dfrac{x}{\|x\|_2}$.

The volume of the unit ball in $\mathbb{R}^n$ is given by:

$$V_n = \frac{\pi^{n/2}}{\Gamma(\frac{n}{2} + 1)}$$

and the volume of the cube $[-1, 1]^n$ is $2^n$. Plot the probability of a vector being accepted in step 2 of this strategy for $n \in \{2, \dots, 20\}$. Do you think this strategy is effective for large $n$ values?

*Hint: We can use* `scipy.special.gamma` *for the Γ function and the function* `scatter` *of Matplotlib to graphically display the values of a sequence with an adapted scale.*

**f.** The following strategy can be shown to generate random vectors on $\mathbb{R}^n$:

1. Generate a vector $x = (x_1, \dots, x_n)$, where each $x_k$ is generated randomly and independently according to the reduced centered normal distribution: $x_k \sim \mathcal{N}(0, 1)$;

2. Return the unit vector $\dfrac{x}{\|x\|_2}$.

Verify for $n = 2$ that this strategy does indeed randomly generate unit vectors according to a uniform distribution by displaying the angle distribution (representing the vectors as complex numbers) for $N = 10^6$ vectors.
*Hint: You can use the function* `numpy.random.standard_normal`.

## EXERCISE 10.6   PERCOLATION (‼)

The goal is to study a percolation model in a porous medium. The medium is modeled by a random matrix of Booleans that determines which sites can be invaded by water and which are impermeable. A matrix percolates if there is a water path from the top row to the bottom row. In the examples in Figure 10.1, the entries of a matrix that can be percolated are colored and the entries that are actually filled with water are in blue.

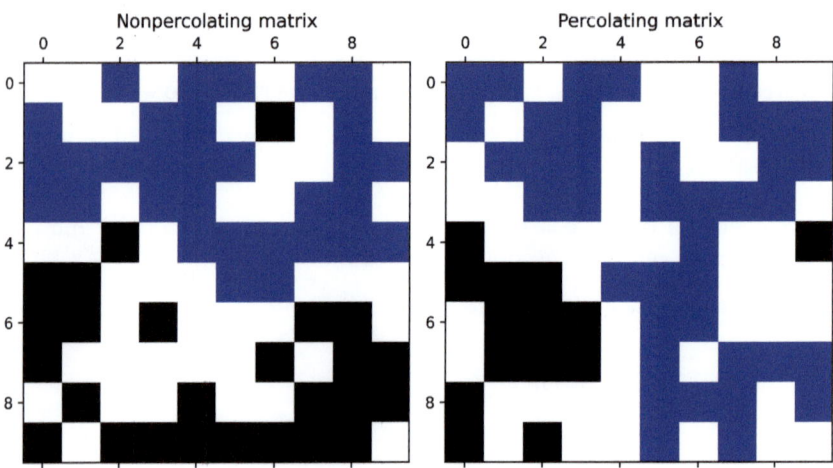

Figure 10.1   The matrix on the left does not percolate while the one on the right does.

**a.** Write a function `generate(n,p)` that generates a matrix of Booleans of size $n \times n$ such that each entry has probability $p$ of being right and $1 - p$ of being wrong.

*Hint: The* `random.binomial` *function of NumPy can be useful.*

**b.** Define a function `fill(isopen)` that for a given Boolean matrix returns another Boolean matrix with the entries invaded by water.

*Hint: Define a Boolean matrix* `isfull` *to store whether an input is filled with water or not, and then define a recursive function* `flow(isopen, isfull, i, j)` *to invade all possible inputs from* $(i, j)$.

**c.** Using Matplotlib, represent the filling of different randomly generated matrices.

**d.** Define a function `percolate(isopen)` to determine whether a Boolean matrix is percolating or not.

**e.** ‼ Calculate the time needed to determine if a matrix of size 50×50 with $p = 0.9$ is percolating or not. Read the documentation of the Numba module to reduce the calculation time by compiling one of the functions: `https://numba.pydata.org/`.

*Hint: The function that is most used is the recursive function, so it is the one that should be optimized when compiling it.*

**f.** By doing statistics, determine the probability that a Boolean random matrix of size $n \times n$ with probability $p$ will percolate. Study this probability as a function of $p$ and $n$.

*Hint: Plot this percolation probability as a function of p for different values of n.*

*Answer: In the limit of n very large, a matrix almost surely percolates if p > 0.592746 and almost never otherwise.*

**g.** ‼‼ The statistics performed in the previous point are a typical example of calculations that can be easily executed in parallel because each case is independent of the others. Parallelize the previous algorithm in such a way as to use all the cores of your processor, for example, with the help of the module `mpi4py`.

*Hint: Using Jupyter Lab to do parallel computing is quite complex to implement, it is better to use the command line to run a script in parallel, for example, for four cores:* `mpirun -n 4 script.py`. *Note that Open MPI or MPICH must be installed on the computer.*

# SOLUTIONS

---

## SOLUTION 10.1   HARMONIC SERIES OF RANDOM SIGN

**a.** It is sufficient to test whether a number generated uniformly in $[0, 1]$ is smaller or larger than $\frac{1}{2}$:

```python
import random
random.seed(1234567)
def sign():
    if random.random() < 1/2:
        return 1
    else:
        return -1
```

**b.** Using the definition:

```python
def simulate(n):
    W = [0]
    for i in range(1,n+1):
        W.append(W[i-1]+sign()/i)
    return W
```

**c.** Taking $n \leq 1\,000$, the sequence $(W_n)_{n\in\mathbb{N}}$ seems to converge as suggested in Figure 10.2, but at each realization to another limit:

```python
n = 1000
plt.figure(figsize=(8,5))
plt.title(f"Simulations of $W_n$ of length {n}")
plt.xlabel("$n$")
for i in range(5):
    plt.plot(simulate(n), label=f"Realization {i}")
plt.legend()
```

**d.** We start by generating a list of $k$ realizations of $W_n$:

```python
n = 1000; k = 10**5
lst = [simulate(n)[-1] for i in range(k)]
```

then we plot the histogram represented in Figure 10.3 to get an idea of the limit:

```python
plt.figure(figsize=(8,5))
plt.title(f"Law of $W_{{{n}}}$ for {k} realizations")
plt.xlabel(f"$W_{{{n}}}$")
plt.hist(lst, bins=100, density=True)
```

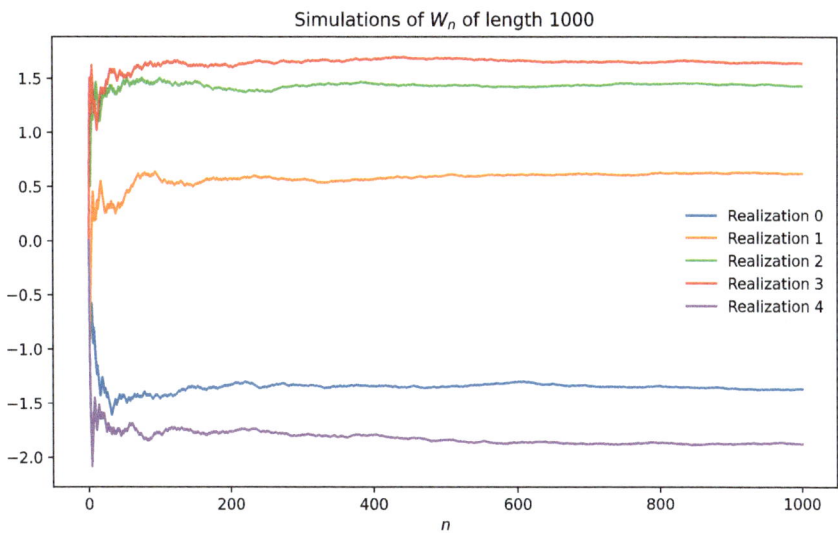

Figure 10.2 Simulations of different realizations of the random sign sequence. The sequence seems to converge, but the value of the limit depends on the realization.

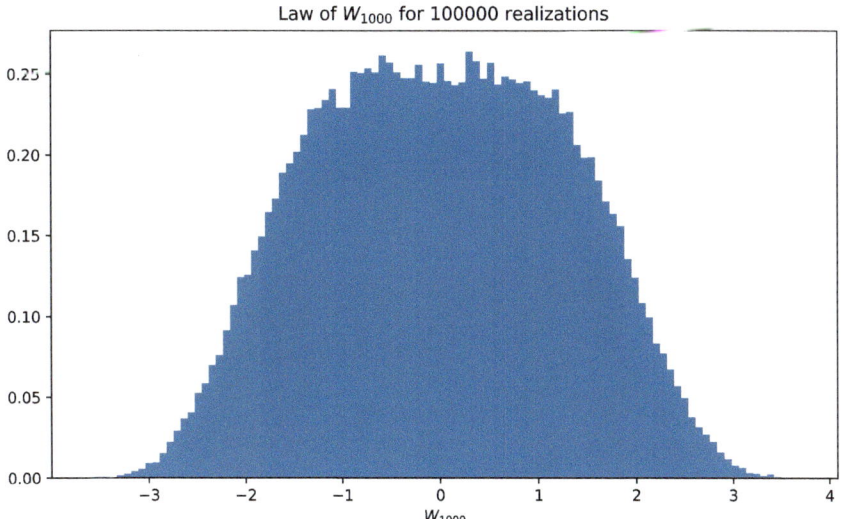

Figure 10.3 Representation of an approximation of the law of $W_{1\,000}$ for $10^5$ realizations.

## SOLUTION 10.2   GAMBLER'S RUIN

**a.** First, we define a function that represents the random variable $X$:

```python
def rand(p):
    if random.random() < p:
        return 1
    else:
        return -1
```

then a function that simulates the gambler's sum:

```python
def simulate(p,k,N):
    S = [k]
    for i in range(N):
        X = rand(p)
        Snew = S[-1] + X
        S.append(Snew)
        if Snew == 0: break
    return S
```

Finally, to represent several trajectories as in Figure 10.4:

```python
p = 1/2; k = 10; N = 1000
plt.figure(figsize=(8,5))
plt.title(f"Processes of length $N$ with p = {p} and k =
 ↪ {k}")
for i in range(5):
    plt.plot(simulate(p,k,N), label=f"Realization {i}")
plt.legend()
```

**b.** The idea is almost identical:

```python
def process(p,k,n):
    S = [k]
    while True:
        X = rand(p)
        Snew = S[-1] + X
        S.append(Snew)
        if Snew == 0 or Snew == n: break
    return S
```

This generates Figure 10.5:

```python
p = 1/2; k = 10; n = 20
plt.figure(figsize=(8,5))
plt.title(f"Process with p = {p}, k = {k} and n = {n}")
for i in range(5):
    plt.plot(process(p,k,n), label=f"Realization {i}")
plt.legend()
```

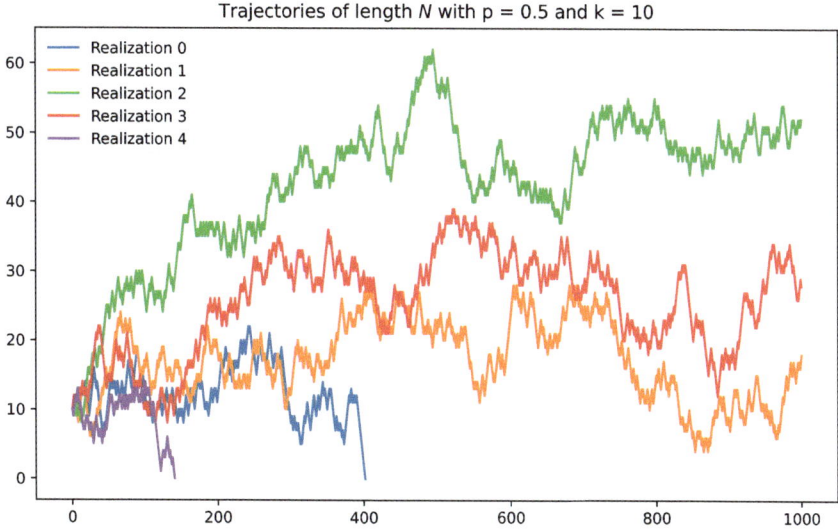

Figure 10.4 Plot of some trajectories in time for the gambler's ruin. On the chosen time window, some realizations already end up at zero, while others have not yet lost everything.

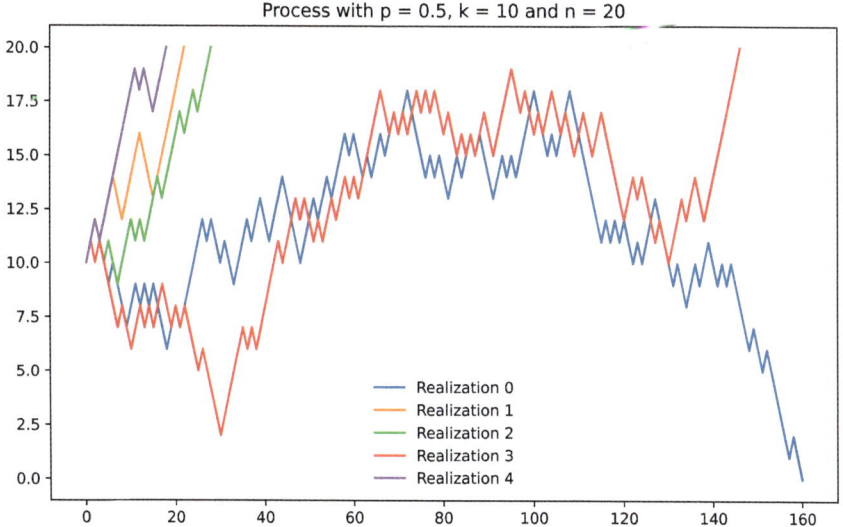

Figure 10.5 Trajectories of the ruin of the player who stops when he has won more than $n = 20$.

**c.** First of all, we define a function that returns the time and place of exit of a process:

```python
def len_last(sim):
    return [len(sim), sim[-1]]
```

Thus using NumPy, we define a function that performs *N* simulations and returns the average of the exit time and the average of the exit location:

```python
def experimental(p,k,n,N):
    lst = np.array([len_last(process(p,k,n)) for _ in
    ↳   range(N)])
    mean_time = np.mean(lst[:,0])
    exit_proba = 1 - np.mean(lst[:,1])/n
    return [mean_time, exit_proba]
```

Then, we define the theoretical values for the same thing:

```python
def theory(p,k,n):
    if p == 0.5:
        mean_time = k*(n-k)
        exit_proba = (n-k)/n
    else:
        q = 1-p
        rho = q/p
        mean_time = n/(p-q)*(1-rho**k)/(1-rho**n) - k/(p-q)
        exit_proba = (rho**k-rho**n)/(1-rho**n)
    return [mean_time,exit_proba]
```

To simulate the whole thing:

```python
k = 10; n = 20; N = 1000;
p = np.linspace(1/1000,1-1/1000,100)
exp = np.array([experimental(v,k,n,N) for v in p])
the = np.array([theory(v,k,n) for v in p])
```

The empirical mean absorption time and probability of ruin follow the theoretical values well, as shown in Figure 10.6:

```python
fig = plt.figure(figsize=(16,6))
plt.suptitle(f"For k = {k} and n = {n}")
# generate the two figures at the same time
for i,title in [(0, "Mean absorption time"), \
                (1, "Probability of ruin")]:
    sub = fig.add_subplot(1,2,1+i)
    sub.set_title(title)
    sub.set_xlabel("p")
    sub.plot(p,exp[:,i], label=f"{N} simulations")
    sub.plot(p,the[:,i], label="Theory")
    sub.legend()
```

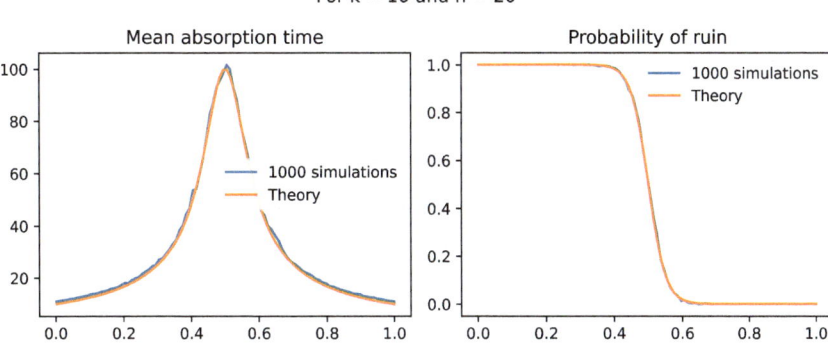

Figure 10.6    Empirical and theoretical probability of the duration of the game as well as of the ruin.

## SOLUTION 10.3    PÓLYA URN

**a.** Simply divide the number of red balls by the total number of balls:

```python
def density(nb_boules):
    r,b = nb_boules
    return r/(r+b)
```

**b.** The function that allows you to move forward in time:

```python
def dist_next_red(r):
    s = len(r) # number of balls at time n+1
    rnext = [0.0]*(s+1)
    rnext[0] = r[0] #  0 unless no red balls
    rnext[s] = r[s-1] # in fact 0 unless at least one white
    ↪    ball
    for i in range(1,s):
        rnext[i] = r[i-1]*(i-1)/(s-1)+r[i]*(s-1-i)/(s-1)
    return rnext
```

and the recursive function to obtain the distribution at time *n*:

```python
def dist_red(r0,b0,n):
    if n==0:
        res = [0.0]*(r0+b0+1)
        res[r0] = 1
        return res
    else:
        return dist_next_red(dist_red(r0,b0,n-1))
```

The conclusion is that when we start with one red ball and one white ball, the distribution of the number of red balls seems uniform over the values from 1 to $n - 1$:

```
dist_red(1,1,20)
```

To illustrate this, it is possible to plot the distributions in different cases, as shown in Figure 10.6:

```
plt.figure(figsize=(8,5))
for i in (0, 1):
    for j,n in enumerate([1, 2, 5, 10]):
        plt.subplot(4, 2, 1+2*j+i)
        plt.bar(np.arange(n+i+2), dist_red(i,1,n))
        plt.xticks(np.arange(n+i+2))
        plt.title(f'dist_red({i}, 1, {n})')
```

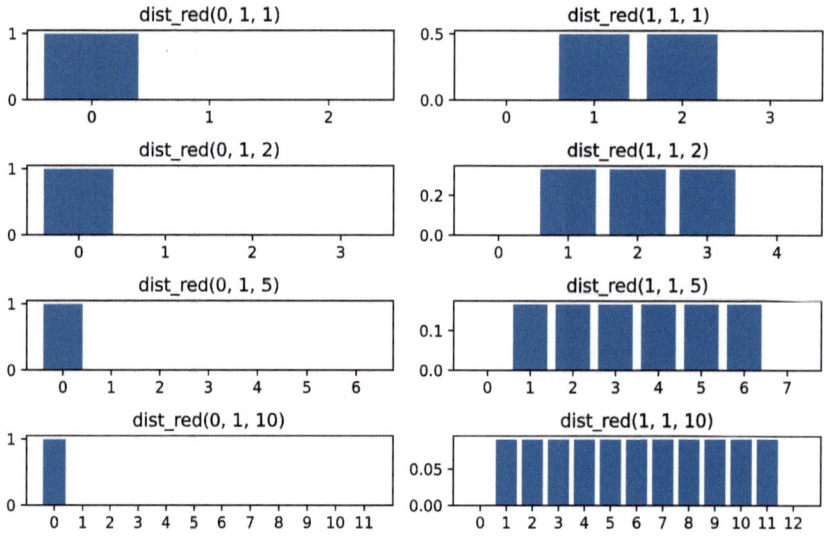

Figure 10.7  Representation of the theoretical distribution of a Pólya urn at times 1, 2, 5, and 10 from a single blue ball or a ball of each color.

**c.** The function to take a step back in time:

```python
def polya_step(r,b):
    u = random.random()
    if u<r/(r+b): # has pick a red ball
        return (r+1,b)
    else:
        return (r,b+1)
```

and the function to simulate a Pólya urn:

```python
def polya(r0,b0,N):
    r,b = r0,b0
    for i in range(N):
        r,b = polya_step(r,b)
    return (r,b)
```

**d.** Just run nbexp Pólya urns and put the results in a list:

```python
def data_rdens_polya(r0,b0,N,nbexp):
    return [density(polya(r0,b0,N)) for _ in range(nbexp)]
```

**e.** The hist function of Matplotlib allows to compare the empirical and theoretical distributions as in Figure 10.8:

```python
r0 = 2; b0 = 3; N = 1000;
data = data_rdens_polya(r0,b0,N,10_000)
theory = np.array(dist_red(r0,b0,N))
plt.figure(figsize=(8,5))
plt.title(f"Distribution of the amount of red balls\n with
    initially {r0} red balls and {b0} white balls")
plt.hist(data, 100, density=True, label="Empirical
    distribution")
plt.plot(np.linspace(0,1,len(theory)), len(theory)*theory,
    label="Theoretical distribution")
plt.legend()
```

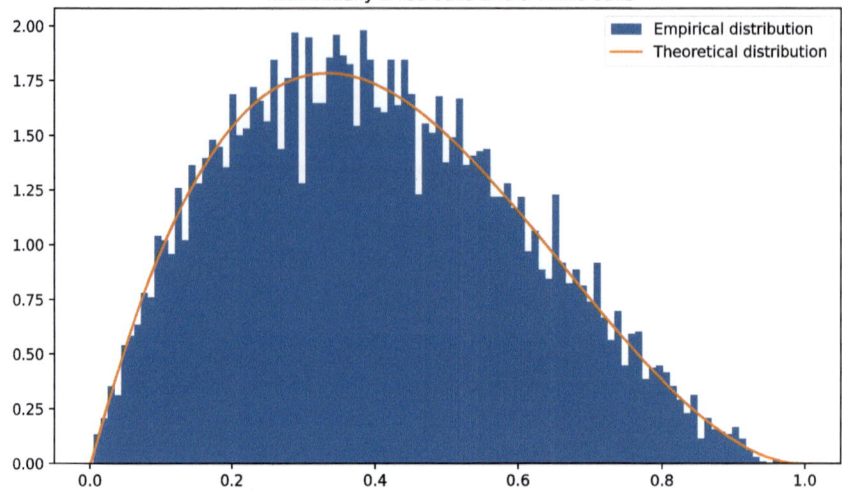

Figure 10.8 Comparison between empirical and theoretical distributions for the proportion of red balls with initially two red and three white balls. The empirical distribution indeed follows the theoretical distribution.

## SOLUTION 10.4 CENTRAL LIMIT THEOREM

**a.** Essentially, it's a matter of copying the formula using NumPy's functions:

```python
def normal_density(x, mu, var):
    return 1 / (np.sqrt(2*np.pi*var)) * np.exp(-(x-mu)**2 /
     ↵ (2*var))
```

**b.** The historic way is to use:

```python
np.random.poisson(lam=2, size=10)
```

The modern way is to use a generator:

```python
rng = np.random.default_rng()
rng.poisson(lam=2, size=10)
```

**c.** This involves using the mean function with the `axis=1` argument to average over the second dimension:

```python
def samples_poisson(lam, N, M):
    sim = rng.poisson(lam=lam, size=(N, M))
    return sim.mean(), sim.std(), sim.mean(axis=1)
```

```
lam = 2 ; N = 10_000 ; Mliste = (10,100,1000)
poisson = [samples_poisson(lam, N, M) for M in Mliste]
```

**d.** Take the previously generated data and plot the histogram and normal distribution with the empirical parameters mean and standard deviation:

```
plt.figure(figsize=(8, 2.5))
for i,result in enumerate(poisson):
        plt.subplot(1, 3, i+1)
        M = Mliste[i]
        mean,std,data = result
        rstd = std/np.sqrt(M) # renormalized standard
          ↵  deviation
        plt.title(f"M = {M}")
        plt.hist(data, bins=50, density=True)
        x = np.linspace(0, 4, 501)
        y = normal_density(x, mean, rstd**2)
        plt.plot(x, y)
        plt.xlim((2-5*rstd,2+5*rstd))
```

The conclusion seems to be that the central limit theorem is satisfied for the Poisson distribution, as shown in Figure 10.9.

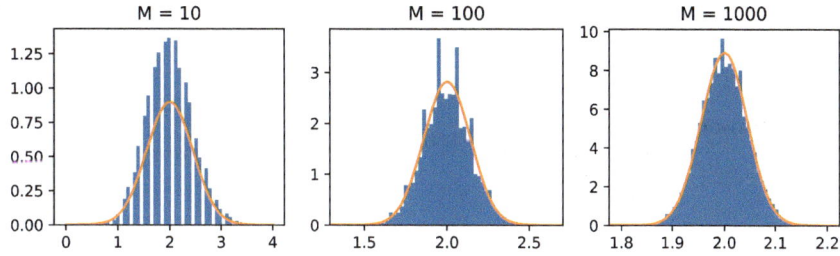

Figure 10.9    Histograms for the Poisson distribution with the corresponding normal distribution for various numbers $M$ of simulated random variables.

**e.** The first step is to generate the samples:

```
def samples_normal(loc, scale, N, M):
    sim = rng.normal(loc=loc, scale=scale, size=(N, M))
    return sim.mean(), sim.std(), sim.mean(axis=1)

loc = 2 ; scale = 1
normal = [samples_normal(loc, scale, N, M) for M in Mliste]
```

The second stage is identical:

```python
plt.figure(figsize=(8, 2.5))
for i,result in enumerate(normal):
        plt.subplot(1, 3, i+1)
        M = Mliste[i]
        plt.title(f"M = {M}")
        mean,std,data = result
        rstd = std/np.sqrt(M)
        plt.hist(data, bins=50, density=True)
        x = np.linspace(0, 4, 501)
        y = normal_density(x, mean, rstd**2)
        plt.plot(x, y)
        plt.xlim((2-5*rstd,2+5*rstd))
```

The conclusion is that the central limit theorem seems to be very well verified for the normal distribution, as represented in Figure 10.10, and this independently of the choice of $M$.

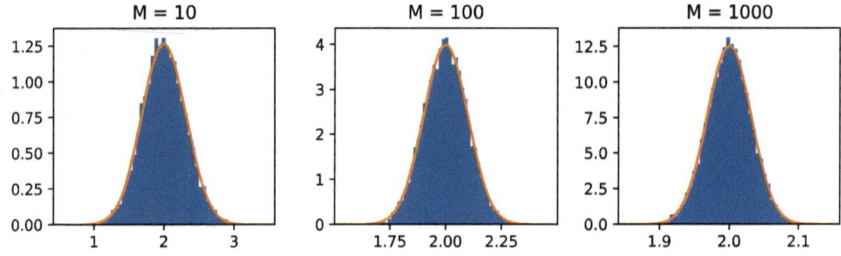

Figure 10.10    Histograms for the normal distribution for various numbers $M$ of simulated random variables.

**f.** The first step is virtually identical:

```python
def echantillons_cauchy(N, M):
    sim = rng.standard_cauchy(size=(N, M))
    return sim.mean(), sim.std(), sim.mean(axis=1)

cauchy = [echantillons_cauchy(N, M) for M in Mliste]
```

To plot histograms with intervals of length 0.1, bins must be set explicitly:

```python
plt.figure(figsize=(8, 2.5))
for i,result in enumerate(cauchy):
        plt.subplot(1, 3, i+1)
        M = Mliste[i]
        plt.title(f"M = {M}")
        mean,std,data = result
```

```
rstd = std/np.sqrt(M)
plt.hist(data, bins=np.arange(-10, 10.1, 0.1),
    ↵ density=True)
x = np.linspace(0, 4, 501)
y = normal_density(x, mean, rstd**2)
plt.plot(x, y)
```

These results shown in Figure 10.11 seem to indicate that the central limit theorem is not valid for Cauchy's law. This is indeed the case, as Cauchy's law has no expectation, so even the weak law of large numbers is invalid.

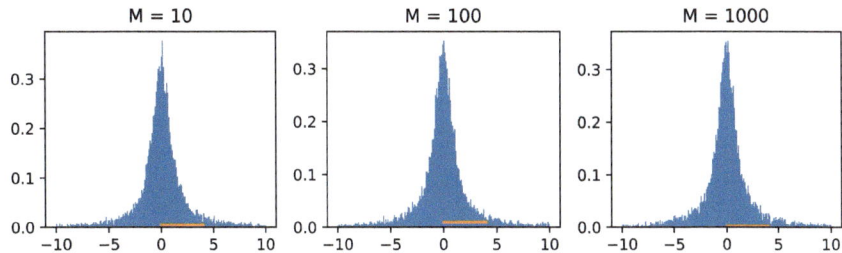

Figure 10.11   Histograms for the Cauchy distribution for various numbers $M$ of simulated random variables. The central limit theorem is not verified.

## SOLUTION 10.5   RANDOM GENERATION OF UNIT VECTORS

**a.** Simply copy the formulas above:

```
def generate_complex(N):
    V = rng.uniform(-1, 1, N) + 1j * rng.uniform(-1, 1, N)
    V /= np.abs(V)
    return V
```

**b.** The easiest way is to use Matplotlib hist function to plot the histogram:

```
plt.hist(np.angle(generate_complex(N=1_000_000)),
    ↵ bins='auto', density=True)
```

The resulting histogram is shown in Figure 10.12. The unit vectors generated in this way are not uniformly distributed. Vectors with angles close to $\frac{\pi}{4} \pm k\frac{\pi}{2}$ are more likely to be generated than vectors with angles close to $0 \pm k\frac{\pi}{2}$ because the corresponding surface area is higher.

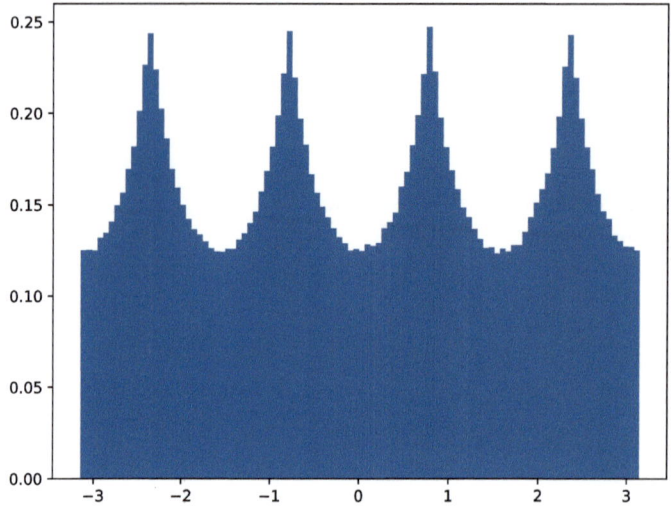

Figure 10.12    Distribution of angles/arguments of complex numbers generated with the algorithm done in **a**.

**c.** With NumPy indexing, the third step is easy:

```python
def generate_complex_monte_carlo(N):
    V = rng.uniform(-1, 1, N) + 1j * rng.uniform(-1, 1, N)
    V_abs = np.abs(V)
    V = V[V_abs <= 1.] / V_abs[V_abs <= 1.]
    return V

plt.hist(np.angle(generate_complex_monte_carlo(N=1_000_000)),
 ↳   bins='auto', density=True)
```

**d.** This is the relationship between the size of the array returned by the previous function and $N$:

```python
N = 1_000_000
abs(generate_complex_monte_carlo(N).size / N - np.pi / 4)
```

**e.** The probability of a vector being accepted is the quotient between the volume of the unit ball $V_n$ and the volume $2^n$ of the cube $[-1, 1]^n$:

```python
from scipy.special import gamma
n = np.arange(2, 41)
volumes = np.pi**(n/2) / gamma(n/2 + 1)
probabilities = volumes / (2**n)
```

To represent this probability in logarithmic scale:

```
plt.scatter(n, probabilities)
plt.yscale('log')
plt.ylim((1e-21, 10))
```

This strategy is clearly not effective: for $n = 40$, you need to generate an average of around $10^{21}$ candidates to accept just one! And this number only increases with higher values of $n$.

**f.** The idea is the same as above:

```
N = 1_000_000
V = rng.standard_normal(N) + 1j * rng.standard_normal(N)
V /= np.abs(V)
```

and for the histogram:

```
plt.hist(np.angle(V), bins='auto', density=True)
```

## SOLUTION 10.6   PERCOLATION (!!)

**a.** We create a random matrix with zeros and ones using NumPy and then convert it to a Boolean matrix:

```
rng = np.random.default_rng(123456)
def generate(n,p):
    # random matrix of 0 and 1 with probability p and (1-p)
    A = rng.binomial(1,p,(n,n))
    # convert to Boolean matrix
    return A==1
```

**b.** First, we define a recursive function allowing to invade all the sites from $(i, j)$:

```
def flow(isopen, isfull, i, j):
    m,n = isopen.shape
    # invalid row
    if i < 0 or i >= m: return
    # invalid column
    if j < 0 or j >= n: return

    # site not open
    if not isopen[i,j]: return
    # site already filled
    if isfull[i,j]: return
    # mark the site as filled
    isfull[i,j] = True

    # invades adjacent sites
```

```
    flow(isopen, isfull, i, j+1)
    flow(isopen, isfull, i, j-1)
    flow(isopen, isfull, i+1, j)
    flow(isopen, isfull, i-1, j)
```

Finally, we just have to initialize an `isfull` matrix and fill it with the elements of the first row:

```
def fill(isopen):
    # generate a matrix of False of the size of isopen
    isfull = np.zeros_like(isopen, dtype=bool)
    # flow starting from each element of the first row
    for j in range(isopen.shape[1]):
        flow(isopen, isfull, 0, j)
    return isfull
```

**c.** A new color bar is defined to color the non-invaded elements black and the invaded ones blue as in Figure 10.13:

```
import matplotlib
cmap = matplotlib.colors.ListedColormap(['w','k','b'])
fig = plt.figure(figsize=(8,5))
for i in range(6):
    sub = fig.add_subplot(2,3,i+1)
    A = generate(10,0.6)
    sub.matshow(1*A+1*fill(A), cmap=cmap)
fig.tight_layout()
```

**d.** To decide if a matrix is percolating, we only need to test if at least one element of the last row is filled:

```
def percolate(isopen):
    # fill the matrix
    isfull = fill(isopen)
    # test if a site of the last row is filled
    for j in range(isfull.shape[1]):
        if isfull[-1,j]: return True
    # if no filled sites on last row
    return False
```

**e.** The typical execution time of:

```
%%timeit
A = generate(50,0.9)
percolate(A)
```

is about 16 ms. The recursive function `flow` being clearly the most used one, let's see what happens if we decide to compile it with Numba:

```
from numba import jit
@jit
def flow(isopen, isfull, i, j):
    m,n = isopen.shape
    # invalid row
    if i < 0 or i >= m: return
    # invalid column
    if j < 0 or j >= n: return

    # site not open
    if not isopen[i,j]: return
    # site already filled
    if isfull[i,j]: return
    # mark the site as filled
    isfull[i,j] = True

    # invades adjacent sites
    flow(isopen, isfull, i, j+1)
    flow(isopen, isfull, i, j-1)
    flow(isopen, isfull, i+1, j)
    flow(isopen, isfull, i-1, j)
```

then:

```
%%timeit
A = generate(50,0.9)
percolate(A)
```

is executed in 590 μs, which is about 30 times faster.

The use of Numba allows to greatly improve the efficiency of functions that cannot be parallelized using NumPy.

**f.** First, a function to determine the percolation probability of a matrix of size $n \times n$ with probability $p$ on average over $N$ realizations:

```
def stats(n,p,N):
    out = np.zeros(N, dtype=bool)
    for i in range(N):
        out[i] = percolate(generate(n,p))
    return np.mean(out)
```

and then to plot as a function of $p$ as in Figure 10.14 :

```
N = 10_000
# list of probabilities
p = np.linspace(0,1,100)
# list of n
n = [10,50,100]
plt.figure(figsize=(8,5))
plt.title(f"Percolation probability for $N = {N}$")
```

```
plt.xlabel(r"$p$")
plt.ylim(-0.1,1.1)
# new curve for each value of n
for i in n:
    data = np.vectorize(lambda p: stats(i,p,N))(p)
    plt.plot(p, data, label=f"n = {i}")
plt.legend()
```

We can clearly see that the larger the *n* is, the more the probability of percolation has a jump. In the limit where *n* is very large, this probability is a jump function: it is a phase transition.

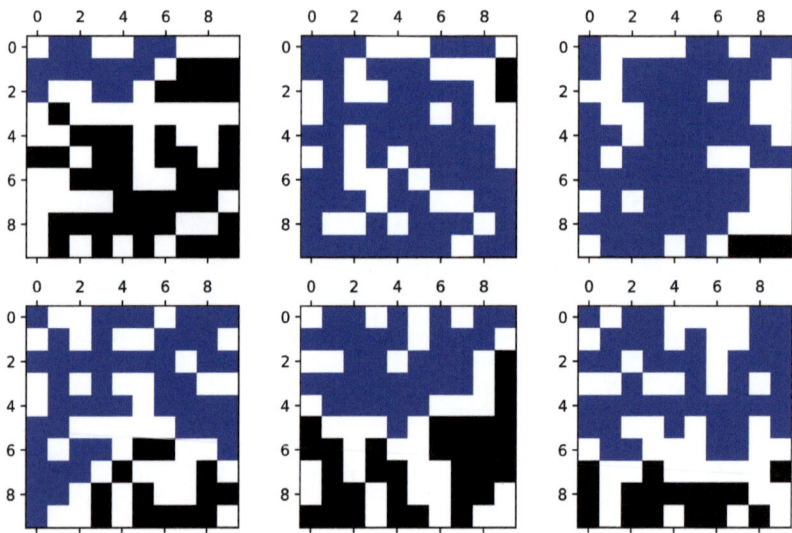

Figure 10.13    In this example, three matrices percolate and three others do not.

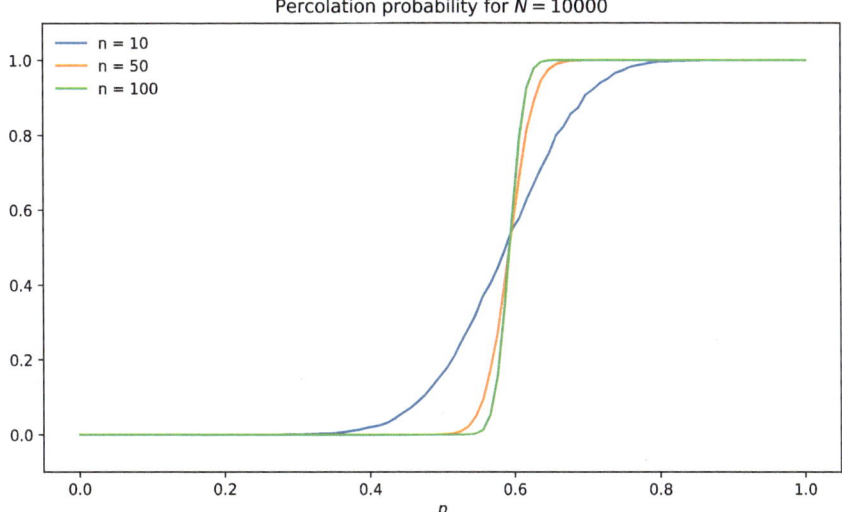

Figure 10.14    Empirical probability of percolation over $N = 10^4$ realizations of a matrix of size $n \times n$ as a function of the filling probability $p$. The larger the $n$ is, the steeper the transition.

# CHAPTER 11

# Differential Equations

The goal is to introduce the basic methods for solving first-order ordinary differential equations of the type:

$$\dot{x}(t) = f(t, x(t)), \qquad\qquad x(0) = x_0,$$

where $f : \mathbb{R}^+ \times \mathbb{R}^n \to \mathbb{R}^n$ is a smooth enough function and $x_0 \in \mathbb{R}^n$ is an initial data. Note that higher-order ordinary differential equations can be put in the previous first-order form.

Concepts covered

- Euler's methods

- Runge-Kutta methods

- nonlinear partial differential equation

- finite differences

- adaptive methods

DOI: 10.1201/9781003565451-11

# EXERCISES

## EXERCISE 11.1   EULER'S METHODS

The simplest idea to approximate an ordinary differential equation is to discretize time with a step $h$ and approximate the time derivative on each interval of length $h$. There are two simple ways to approximate the time derivative. The first is the forward finite difference approximation:

$$\dot{x}(t) \approx \frac{x(t+h) - x(t)}{h},$$

the second, the backward finite difference:

$$\dot{x}(t) \approx \frac{x(t) - x(t-h)}{h}.$$

The unknowns being the evaluations of the solution $x$ at times $t_i = ih$ for $i \geq 0$, i.e., $x_i = x(t_i)$. The differential equation can thus be approximated using forward finite differences by:

$$\frac{x_{i+1} - x_i}{t_{i+1} - t_i} = f(t_i, x_i),$$

which gives the explicit Euler formula:

$$x_{i+1} = x_i + (t_{i+1} - t_i)f(t_i, x_i).$$

With the backward finite difference approximation, we obtain the implicit Euler method (also called backward Euler method):

$$x_i = x_{i-1} + (t_i - t_{i-1})f(t_i, x_i).$$

On the one hand, the explicit Euler formula allows to compute directly all $x_i$ by recurrence knowing $x_0$. On the other hand, the implicit Euler formula requires at each time step the resolution of a nonlinear equation for $x_i$, for example, with the Newton's method.

**a.** Write a function `euler_explicit(f,x0,t)` that given an initial data x0 returns the values $x_0, x_1, ..., x_m$ computed with the explicit Euler method at times $(t_i)_{i=0}^m$ represented by the vector t.

**b.** Use the explicit Euler method to solve the differential equation:

$$\dot{x}(t) + x(t) = \sin(t), \quad x(0) = 1,$$

for $t \in [0, 10]$. Compare the results with the exact solution:

$$x(t) = \frac{1}{2}\left(\sin(t) - \cos(t) + 3e^{-t}\right),$$

for different time discretizations.

**c.** Solve the previous problem with the implicit Euler method.
*Hint: Since the previous equation is linear, we can actually make the implicit Euler method explicit by solving the implicit equation by hand.*

**d.** !! Define a function `euler_implicit(f, Dxf, x0, t)` implementing the implicit Euler method for nonlinear equations. Note that to solve the nonlinear problem with Newton's method, the derivative of $f$ according to $x$ is required.
*Hint: It is also possible to use the root finding algorithm* `optimize.fsolve` *from SciPy that does not require to know the derivative of $f$.*

**e.** ! Use the previous methods to find an approximate solution of the system:

$$\dot{x}(t) + \cos(y(t)) = \sin(t), \qquad\qquad x(0) = 1,$$
$$\dot{y}(t) + \cos(x(t)) = 0, \qquad\qquad y(0) = 0.$$

## EXERCISE 11.2   RUNGE-KUTTA METHODS

The purpose of this exercise is to introduce a class of methods more accurate than Euler's methods for solving ordinary differential equations. Instead of doing a first order approximation in $h$ the idea is to do a higher-order approximation of the derivative.

The basic idea is to construct a sequence $x_i$ giving an approximation of the solution of $\dot{x}(t) = f(t, x)$ at time $t_i$ for $i \in \mathbb{N}$. This sequence is defined by:

$$x_{i+1} = x_i + M(t_i, x_i, t_{i+1} - t_i),$$

for a certain function $M$ called method. For example, for the explicit Euler method, the function $M$ is given by:

$$M(t, x, h) = hf(t, x).$$

A Runge-Kutta method of order two is given by:

$$M(t, x, h) = hf\left(t + \frac{h}{2}, x + \frac{h}{2}f(t, x)\right).$$

A Runge-Kutta method of order four is given by:

$$M(t, x, h) = \frac{h}{6}(k_1 + 2k_2 + 2k_3 + k_4),$$

where

$$k_1 = f(t, x),$$
$$k_2 = f\left(t + \frac{h}{2}, x + \frac{h}{2}k_1\right),$$
$$k_3 = f\left(t + \frac{h}{2}, x + \frac{h}{2}k_2\right),$$
$$k_4 = f(t + h, x + hk_3).$$

Note that more generally, a Runge-Kutta method of order $s$ is given by:

$$M(t, \boldsymbol{x}, h) = h \sum_{i=1}^{s} b_i \boldsymbol{k}_i,$$

where

$$\boldsymbol{k}_1 = f(t, \boldsymbol{x}),$$
$$\boldsymbol{k}_2 = f(t + c_2 h, \boldsymbol{x} + h a_{21} \boldsymbol{k}_1),$$
$$\boldsymbol{k}_3 = f(t + c_3 h, \boldsymbol{x} + h(a_{31} \boldsymbol{k}_1 + a_{32} \boldsymbol{k}_2)),$$
$$\vdots$$
$$\boldsymbol{k}_s = f(t + c_s h, \boldsymbol{x} + h(a_{s1} \boldsymbol{k}_1 + a_{s2} \boldsymbol{k}_2 + \cdots + a_{s,s-1} \boldsymbol{k}_{s-1})).$$

The coefficients $a_{ij}$ (for $1 \le j < i \le s$), $c_i$ (for $2 \le i \le s$), and $b_i$ (for $1 \le i \le s$) are often represented in a Butcher table:

$$
\begin{array}{c|ccccc}
0 & & & & & \\
c_2 & a_{21} & & & & \\
c_3 & a_{31} & a_{32} & & & \\
\vdots & \vdots & & \ddots & & \\
c_s & a_{s1} & a_{s2} & \cdots & a_{s,s-1} & \\
\hline
 & b_1 & b_2 & \cdots & b_{s-1} & b_s
\end{array}
$$

For example, the Butcher array from the previous method of order two is:

$$
\begin{array}{c|cc}
0 & & \\
\frac{1}{2} & \frac{1}{2} & \\
\hline
 & 0 & 1
\end{array}
$$

and that of the fourth order method:

$$
\begin{array}{c|cccc}
0 & & & & \\
\frac{1}{2} & \frac{1}{2} & & & \\
\frac{1}{2} & 0 & \frac{1}{2} & & \\
1 & 0 & 0 & 1 & \\
\hline
 & \frac{1}{6} & \frac{1}{3} & \frac{1}{3} & \frac{1}{6}
\end{array}
$$

**a.** Define a function `integrate(f, x0, t, M)` which for a given list of times $(t_i)_{i=0}^{N}$ returns the corresponding values $\boldsymbol{x}_0, \boldsymbol{x}_1, \ldots, \boldsymbol{x}_N$ with method $M$.

**b.** Implement the functions `M(f,t,x,h)` for the explicit Euler method and the Runge-Kutta method of order two. Compare the two methods.

**c.** Implement the function M(f,t,x,h) for the Runge-Kutta method of order four. Compare with the second-order method.

## EXERCISE 11.3   MOVEMENT OF A PLANET

The goal is to simulate the two-dimensional motion of a planet orbiting around a fixed star. The star is supposed to be fixed at the origin and the position of the planet in the plane is described by the vector $x \in \mathbb{R}^2$. The star is supposed to interact with the planet with the potential:

$$V(x) = \frac{1}{\alpha}|x|^\alpha,$$

for a certain $\alpha \in \mathbb{R}$, where $|x|$ denotes the euclidean norm of the vector $x$. Note that the gravitational potential corresponds to $\alpha = -1$. The equation of the planet in this force field is given by:

$$\ddot{x} = -\nabla V(x) = -x|x|^{\alpha-2}.$$

**a.** Rewrite the second-order differential equation as a first-order differential equation for $x$ and $p = \dot{x}$.

**b.** Implement the function f(t,xp) corresponding to the equation found in the previous point.

**c.** Using the Runge-Kutta method of fourth order, solve the differential equation for different initial data and different values of $\alpha$ and plot the $x(t)$ trajectories in the plane. Interpret the results and explain in particular why the cases $\alpha = -1$ and $\alpha = 2$ are different from the others.

## EXERCISE 11.4   LORENZ ATTRACTOR

The Lorenz model is a system of three coupled differential equations of the form

$$\dot{x} = \sigma(y - x),$$
$$\dot{y} = x(\rho - z) - y,$$
$$\dot{z} = xy - \beta z,$$

where $\rho, \sigma, \beta$ are three real parameters. This is a very simplified model of coupling between the atmosphere and the ocean proposed in 1963 by Edward Lorenz.

**a.** Write mathematically the expression of the function $f : \mathbb{R} \times \mathbb{R}^3$ allowing to put the Lorenz system in the form

$$\dot{x} = f(t, x),$$

where $x$ is the vector $(x, y, z)$. Implement a function f(t,x,rho,sigma,beta) corresponding to the function $f$.

**b.** Write a function `plot_lorenz(rho,sigma,beta)` which for given parameters $\rho, \sigma, \beta$, plots the trajectory $(x(t), z(t))$ for $t \in [0, 20]$ from the initial data $x_0 = (x_0, y_0, z_0) = (1, 1, 1)$. Use, for example, the Runge-Kutta method of order four with a time step $\Delta t = 0.001$. Test with $\sigma = 10$, $\beta = 8/3$, and the values $\rho = 10, 15, 20, 25$, and describe what is observed.

**c.** Using SymPy, determine the stationary solutions according to the parameters $\rho, \sigma, \beta$, i.e., the solutions of $f(t, x) = 0$ for all $t > 0$. Interpret the previous graphs in the light of this.

## EXERCISE 11.5   CUBIC WAVE EQUATION (!!)

The goal is to solve numerically the nonlinear wave equation on $\mathbb{R}$:

$$-\frac{\partial^2 u}{\partial t^2} + \frac{\partial^2 u}{\partial x^2} = u^3, \qquad u(0, \cdot) = u_0, \qquad \frac{\partial u}{\partial t}(0, \cdot) = v_0,$$

for $u : \mathbb{R}^+ \times \mathbb{R} \to \mathbb{R}$ with $u_0, v_0 : \mathbb{R} \to \mathbb{R}$ two given functions.

**Remark:** The properties of this seemingly simple equation are very poorly understood mathematically, see the following research article for more details: doi:10.2140/apde.2012.5.411.

**a.** Rewrite the previous equation as two first-order equations in time for $u$ and $v = \frac{\partial u}{\partial t}$.

**b.** By approximating the second derivative in space by finite differences as in Exercise 9.4, show that the equation can be approximated as follows:

$$\frac{\partial u_n}{\partial t} = v_n, \qquad\qquad u_n(0) = u_0(x_n),$$

$$\frac{\partial v_n}{\partial t} = \frac{u_{n-1} - 2u_n + u_{n+1}}{h^2} - u_n^3, \qquad v_n(0) = v_0(x_n),$$

where $(x_n)_{n=0}^N$ denotes $N + 1$ evenly spaced points from $h$ in the interval $[-L, L]$ and $u_n(t) = u(t, x_n)$ and $v_n(t) = v(t, x_n)$. For the conditions at the boundary of the domain, i.e., when $n = 0$ or $n = N$, we take:

$$\frac{\partial v_0}{\partial t} = 0, \qquad \frac{\partial v_N}{\partial t} = 0.$$

**c.** Determine the function $f : \mathbb{R}^{2N+2} \to \mathbb{R}^{2N+2}$ allowing to put the previous approximation in the form $\dot{u} = f(t, u)$ for $u = (u, v)$ and implement this function.

**d.** Solve the differential equation given by $\dot{u} = f(t, u)$, for example, with the fourth-order Runge-Kutta method. A good choice of parameters is $L = 100$, $N = 1\,000$ and for the initial data $u_0(x) = e^{-x^2}$ and $v_0(x) = 0$. The speed of propagation of the wave is one and, therefore, after a time greater than $L$, the wave leaves the box $[-L, L]$ and no longer corresponds to a good approximation of the initial equation.

**e.** Using the `animation` module of Matplotlib, make a video showing the evolution of the wave as a function of time.
*Hint: Use, for example, the* `FFMpegWriter` *function.*

## EXERCISE 11.6 BOGACKI-SHAMPINE METHODS (!!!)

By combining two Runge-Kutta methods of different orders (for example, (2,3) or (4,5)), one will obtain an empirical estimate of the error over a time step. Using this error estimate, it is possible to adapt the time step, either by increasing or decreasing it, and thus adapt to the equation.

For a Runge-Kutta method of order $s$, an internal method of lower order (usually $s-1$) is given by:

$$M^*(t, x, h) = h \sum_{i=1}^{s} b_i^* k_i,$$

where the $k_i$ are identical to those of the $s$ order method. An estimate of the error is then given by:

$$E(t, x, h) = M(t, x, h) - M^*(t, x, h) = h \sum_{i=1}^{s} (b_i - b_i^*) k_i.$$

Such a method is given by an extended Butcher table:

| | | | | | |
|---|---|---|---|---|---|
| 0 | | | | | |
| $c_2$ | $a_{21}$ | | | | |
| $c_3$ | $a_{31}$ | $a_{32}$ | | | |
| $\vdots$ | $\vdots$ | | $\ddots$ | | |
| $c_s$ | $a_{s1}$ | $a_{s2}$ | $\cdots$ | $a_{s,s-1}$ | |
| | $b_1$ | $b_2$ | $\cdots$ | $b_{s-1}$ | $b_s$ |
| | $b_1^*$ | $b_2^*$ | $\cdots$ | $b_{s-1}^*$ | $b_s^*$ |

**a.** Implement the Bogacki-Shampine method of order (4,5). The original article is available at doi:10.1016/0898-1221(96)00141-1.
*Hint: The coefficients of the Butcher tables are implemented in the* nodepy *package, whose documentation is available at the address:* https://nodepy.readthedocs.io/. *The name of the method in this package is "BS5".*

# SOLUTIONS

---

## SOLUTION 11.1   EULER'S METHODS

**a.** First of all, load NumPy and Matplotlib:

```python
import numpy as np
import matplotlib.pyplot as plt
```

The following implementation allows to handle systems of equations, hence **x** is a matrix:

```python
def euler_explicit(f, x0, t):
    # initialize the vector solution
    x = np.zeros((len(t),len(x0)))
    # initial data
    x[0] = x0
    # time stepping
    for i in range(len(t)-1):
        x[i+1] = x[i] + (t[i+1]-t[i])*f(t[i],x[i])
    return x
```

**b.** For this equation, the function $f$ is given by:

$$f(t, x) = \sin(t) - x.$$

The finer the discretization, the closer the approximated solution is to the exact solution as in Figure 11.1:

```python
# define problem data
f = lambda t,x : np.sin(t) - x
x0 = np.array([1])
# figure
plt.figure(figsize=(8,5))
plt.title(r"Solution of $\dot{x} + x = \sin(t)$ using the
    explicit Euler method")
# various discretizations
for N in [10,20,50,100]:
    t = np.linspace(0,10,N)
    sol = euler_explicit(f, x0, t)
    plt.plot(t, sol, label=f"solution with {N} points")
exact = (np.sin(t) - np.cos(t) + 3*np.exp(-t))/2
plt.plot(t, exact, label="exact solution")
plt.legend()
```

**c.** The implicit Euler iterations are given by:

$$x_i = x_{i-1} + (t_i - t_{i-1})(\sin(t_i) - x_i),$$

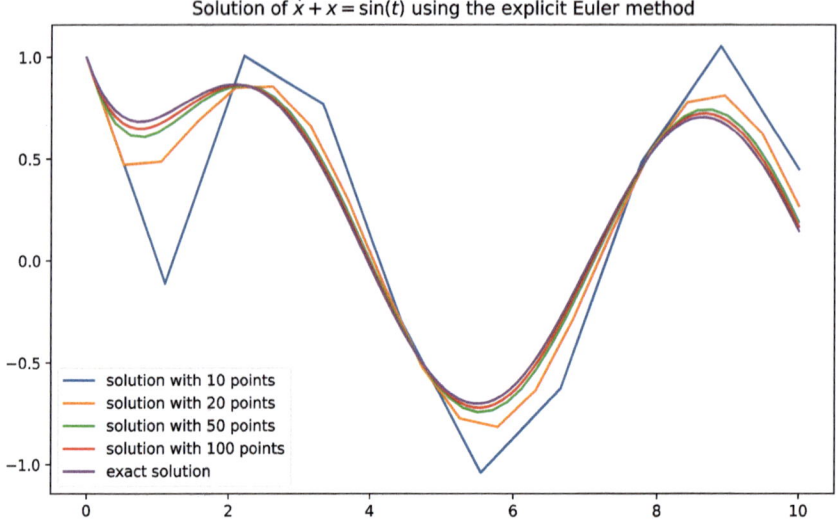

**Figure 11.1** Approximation of the solution of the differential equation $\dot{x}+x = \sin$ by the explicit Euler method for different discretizations.

*i.e.*, explicitly by:

$$x_i = \frac{1}{1 + t_i - t_{i-1}}\left(x_{i-1} + (t_i - t_{i-1})\sin(t_i)\right).$$

It is then sufficient to implement this recurrence:

```python
def solve_eq(t):
    # define problem data
    f = lambda x,t : np.sin(t) - x
    x0 = 1
    # initialize the vector solution
    x = np.zeros(len(t))
    # initial data
    x[0] = x0
    # time loop
    for i in range(1,len(t)):
        x[i] = (x[i-1] + (t[i]-t[i-1])*np.sin(t[i])) / (1 +
        ↵  t[i] - t[i-1])
    return x
```

By testing it, we notice in Figure 11.2 that the implicit Euler method is hardly more accurate than the explicit one for large time steps:

```
# figure
plt.figure(figsize=(8,5))
plt.title(r"Solution of $\dot{x} + x = \sin(t)$ using the
  ↳ implicit Euler method")
# various discretizations
for N in [10,20,50,100]:
    t = np.linspace(0,10,N)
    sol = solve_eq(t)
    plt.plot(t, sol, label=f"solution with {N} points")
exact = (np.sin(t) - np.cos(t) + 3*np.exp(-t))/2
plt.plot(t, exact, label="exact solution")
plt.legend()
```

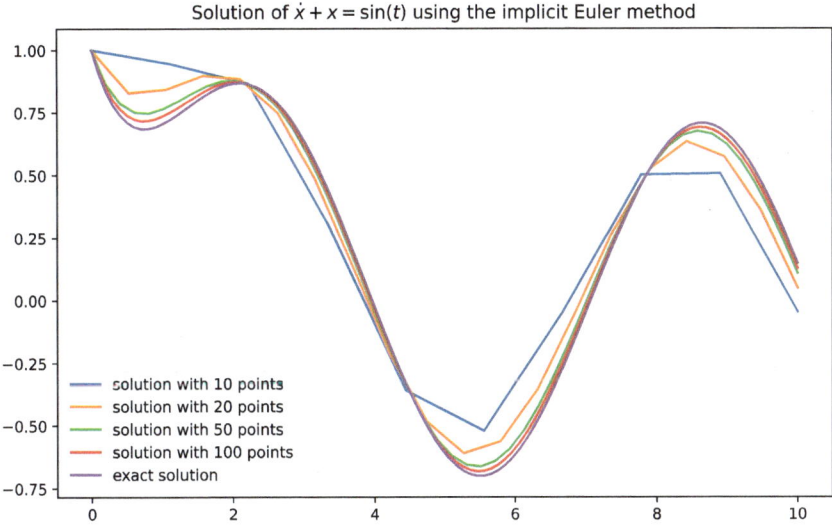

Figure 11.2   Solving a linear differential equation by the implicit Euler method, thus, not requiring the solution of an algebraic equation.

**d.** First, we recall Newton's algorithm from Exercise 9.2 for solving a nonlinear equation:

```
def newton(F, DF, x0, eps=1e-12, N=10000):
    x = x0.copy()
    for i in range(N):
        # calculate F(x) and DF(x)
        Fx = F(x)
        DFx = DF(x)
        # test if the precision is sufficient
        if np.linalg.norm(Fx) < eps:
```

```
        return x
    # test that the derivative is not too small
    if np.linalg.norm(DFx) < eps:
        raise Exception(f"Derivative   |DF| =
        ↪  {np.linalg.norm(DFx)} too small")
    # solve d = DFx^{-1} Fx then Newton's iteration
    x -= np.linalg.solve(DFx, Fx)
# if the loop ends, one has not converged (yet)
raise Exception(f"The error after {N} iterations is
↪  {np.linalg.norm(Fx)} > {eps}")
```

The nonlinear equation to solve is given by $F(x_i) = 0$ for $F : \mathbb{R}^n \to \mathbb{R}^n$ defined by:

$$F(x_i) = x_i - x_{i-1} - (t_i - t_{i-1})f(t_i, x_i).$$

The derivative of $F$ is therefore given by:

$$F'(x) = I_n - (t_i - t_{i-1})D_x f(t_i, x_i),$$

where $I_n$ denotes the identity matrix of size $n \times n$ and $D_x f$ the derivative of $f$ with respect to $x$ only. Then, the implicit Euler method is given by:

```
def euler_implicit(f, Dxf, x0, t):
    # initialize the vector solution
    x = np.zeros((len(t),len(x0)))
    # intitial data
    x[0] = x0
    # time loop
    for i in range(1,len(t)):
        # function whose zero must be found
        F = lambda xi: xi - x[i-1] - (t[i]-t[i-1])*f(t[i],xi)
        # derivative of this function
        DF = lambda xi: np.identity(len(x0)) -
        ↪  (t[i]-t[i-1])*Dxf(t[i],xi)
        # solution using Newton's method
        x[i] = newton(F, DF, x[i-1])
    return x
```

**e.** The function $f$ is given by:

$$f(t, x, y) = \begin{pmatrix} \sin(t) - \cos(y) \\ -\cos(x) \end{pmatrix},$$

and so its derivative with respect to $x = (x, y)$ is given by:

$$D_x f(t, x, y) = \begin{pmatrix} 0 & \sin(y) \\ \sin(x) & 0 \end{pmatrix}.$$

So using the functions defined above, we obtain Figure 11.3:

```
# define problem data
f = lambda t,x : np.array([np.sin(t)-np.cos(x[1]),
  ↵  -np.cos(x[0])])
Dxf = lambda t,x : np.array([[0,
  ↵  np.sin(x[1])],[np.sin(x[0]),0]])
x0 = np.array([1,0])
# figure
plt.figure(figsize=(8,5))
plt.title(r"Solution of $\dot{x}+\cos(y)=\sin(t)$ and
  ↵  $\dot{y}+\cos(x)=0$")
# resolution
t = np.linspace(0,10,101)
sol = euler_explicit(f, x0, t)
plt.plot(t, sol[:,0], label=r"explicit Euler: $x(t)$")
plt.plot(t, sol[:,1], label=r"explicit Euler: $y(t)$")
sol = euler_implicit(f, Dxf, x0, t)
plt.plot(t, sol[:,0], label=r"implicit Euler: $x(t)$")
plt.plot(t, sol[:,1], label=r"implicit Euler: $y(t)$")
plt.legend()
```

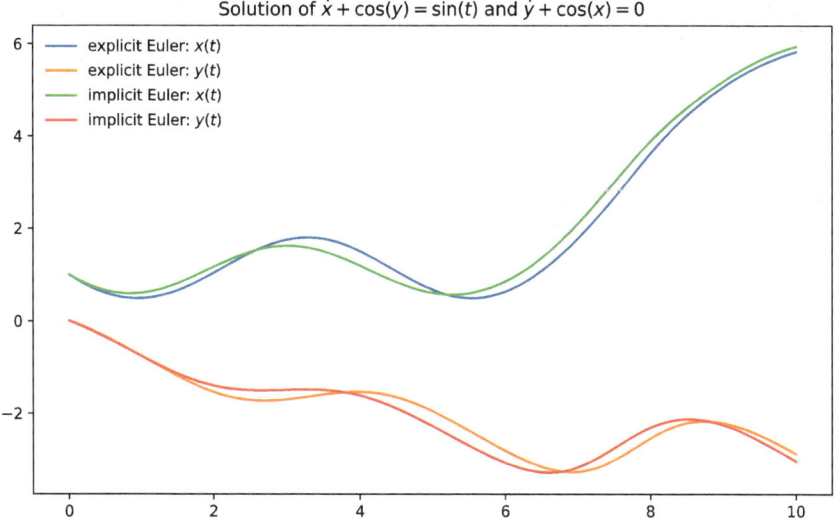

**Figure 11.3**  Comparison of the explicit and implicit Euler methods for a system of differential equations with a time step of $h = 0.1$.

## SOLUTION 11.2  RUNGE-KUTTA METHODS

**a.** The code is almost identical to the one used for the explicit Euler method:

```python
def integrate(f,x0,t,M):
    # initialize vector solution
    x = np.zeros((len(t),) + x0.shape)
    # initial data
    x[0] = x0
    # time stepping
    for i in range(len(t)-1):
        x[i+1] = x[i] + M(f, t[i], x[i], t[i+1]-t[i])
    return x
```

This implementation allows x to also be an array with more than one dimension.

**b.** The method for the explicit Euler method is:

```python
def euler(f,t,x,h):
    return h*f(t,x)
```

and that for Runge-Kutta of order two:

```python
def rk2(f,t,x,h):
    return h*f(t+h/2, x+h/2*f(t,x))
```

As shown in Figure 11.4, the second-order method is quite a bit more accurate than the explicit Euler method which is of order one:

```python
f = lambda t,x: np.sin(t) - x
x0 = np.array([1])
t = np.linspace(0,10,21)
plt.figure(figsize=(8,5))
plt.title("Comparison between explicit Euler and RK2")
for M in [euler,rk2]:
    sol = integrate(f,x0,t,M)
    plt.plot(t,sol,label=M.__name__)
t = np.linspace(0,10,200)
exact = (np.sin(t) - np.cos(t) + 3*np.exp(-t))/2
plt.plot(t,exact,label="exact")
plt.legend()
```

**c.** It is enough to calculate the $k_i$ then to sum:

```python
def rk4(f,t,x,h):
    k1 = f(t,x)
    k2 = f(t+h/2, x+h/2*k1)
    k3 = f(t+h/2, x+h/2*k2)
    k4 = f(t+h, x+h*k3)
    return h/6*(k1+2*k2+2*k3+k4)
```

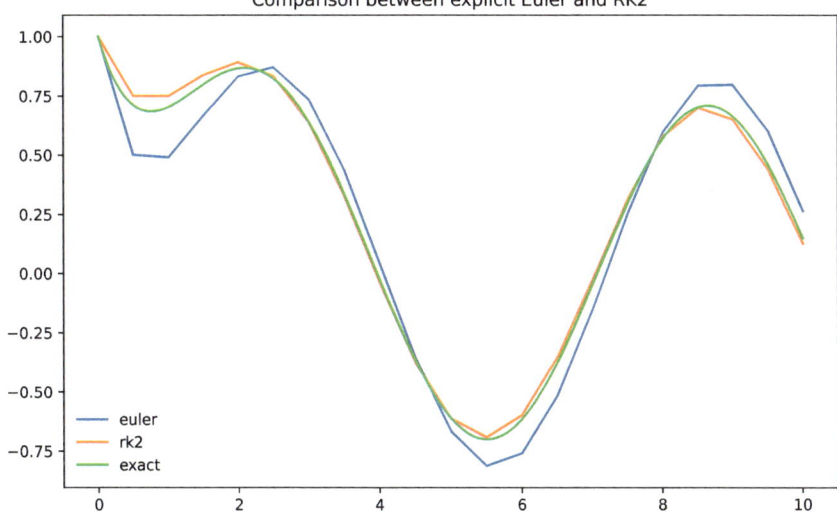

Figure 11.4 As expected the Runge-Kutta method of order two is more accurate than the explicit Euler method for the same time step $h = 0.5$.

The fourth-order method allows to be very close to the exact solution even with a large time step, as shown in Figure 11.5:

```
f = lambda t,x: np.sin(t) - x
x0 = np.array([1])
t = np.linspace(0,10,10)
plt.figure(figsize=(8,5))
plt.title("Comparison between RK2 and RK4")
for M in [rk2,rk4]:
    sol = integrate(f,x0,t,M)
    plt.plot(t,sol,label=M.__name__)
t = np.linspace(0,10,200)
exact = (np.sin(t) - np.cos(t) + 3*np.exp(-t))/2
plt.plot(t,exact,label="exact")
plt.legend()
```

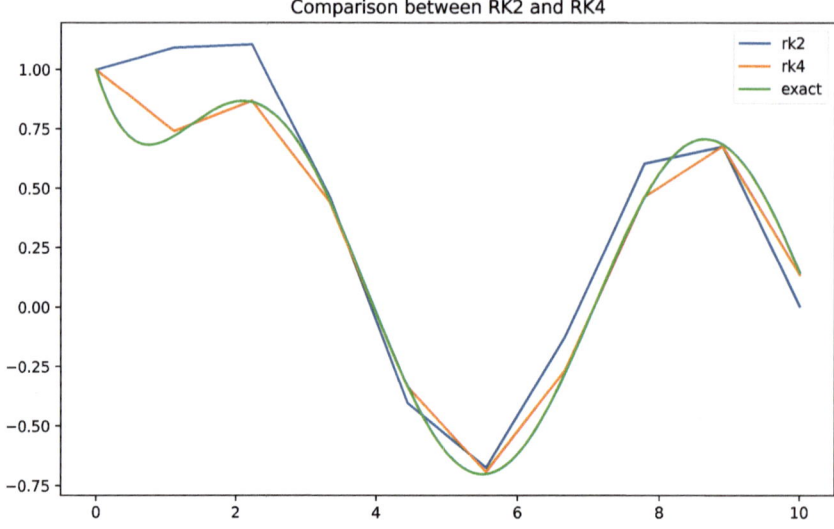

**Figure 11.5** The fourth-order Runge-Kutta method provides a practically indistinguishable approximation to the exact solution even for $h = 1$.

## SOLUTION 11.3   MOVEMENT OF A PLANET

**a.** The equation is rewritten as a first-order system of dimension four:

$$\dot{x} = p, \qquad\qquad \dot{p} = -x|x|^{\alpha-2}.$$

**b.** The vector xp represents the concatenation of the vectors $x$ and $p$, so the indices 0 and 1 correspond to $x$ and the indices 2 and 3 to $p$:

```python
def f(t,xp):
    # result dxp=(\dot{x},\dot{p})
    dxp = xp.copy()
    # derivative of x
    dxp[0:2] = xp[2:4]
    # derivative of p
    dxp[2:4] = -xp[0:2]*np.linalg.norm(xp[0:2])**(alpha-2)
    return dxp
```

**c.** For $\alpha = -1$ and $\alpha = 2$, the trajectories are either ellipses, circles, or hyperbolas. For example, with $x_0 \propto (1,0)$ and $p_0 = (0,1)$, we obtain Figure 11.6:

```python
fig = plt.figure(figsize=(12,5))
fig.suptitle(r'Closed bounded orbits for $\alpha=-1$ and
↪    $\alpha=2$')
# alpha=-1 and alpha=2
```

```
for i,alpha in enumerate([-1,2]):
    sub = fig.add_subplot(1,2,i+1)
    sub.set_title(f"$\\alpha = {alpha}$")
    for s in [0.2,0.5,0.8,1,1.2,1.5,2]:
        x0 = np.array([s,0,0,1])
        # adapt time intervals on the cases
        if alpha==-1:
            t = np.linspace(0,10*s**2,10000)
        elif alpha==2:
            t = np.linspace(0,10,10000)
        sol = integrate(f, x0, t, rk4)
        sub.plot(sol[:,0], sol[:,1], label=f"$x_0={s}$")
    sub.set_xlim([-2,2])
    sub.set_ylim([-2,2])
    sub.set_aspect('equal')
    sub.legend()
```

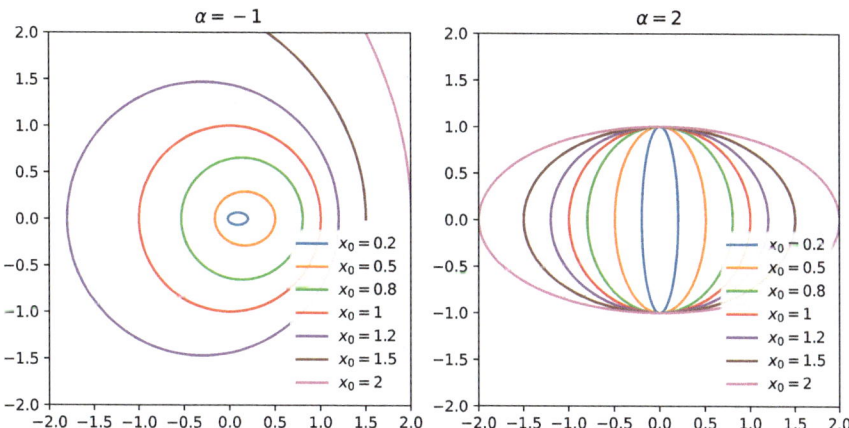

Figure 11.6   For the gravitational potential ($\alpha = -1$) and the harmonic potential ($\alpha = 2$), the trajectories are closed.

For other values of $\alpha$, this gives Figure 11.7:

```
fig = plt.figure(figsize=(12,12))
fig.suptitle(r'Bounded unclosed orbits for $\alpha\neq-1$ or
    ↪ $\alpha\neq2$')

alpha = -1.5
sub = fig.add_subplot(2,2,1)
sub.set_title(f"$\\alpha = {alpha}$")
x0 = np.array([1.1,0,0,1])
```

```
t = np.linspace(0,500,10000)
sol = integrate(f, x0, t, rk4)
sub.plot(sol[:,0], sol[:,1])
sub.set_xlim([-3,3])
sub.set_ylim([-3,3])
sub.set_aspect('equal')

alpha = -0.5
sub = fig.add_subplot(2,2,2)
sub.set_title(f"$\\alpha = {alpha}$")
x0 = np.array([2,0,0,1])
t = np.linspace(0,500,10000)
sol = integrate(f, x0, t, rk4)
sub.plot(sol[:,0], sol[:,1])
sub.set_xlim([-4,4])
sub.set_ylim([-4,4])
sub.set_aspect('equal')

alpha = 1.5
sub = fig.add_subplot(2,2,3)
sub.set_title(f"$\\alpha = {alpha}$")
x0 = np.array([0.1,0,0,1])
t = np.linspace(0,95,10000)
sol = integrate(f, x0, t, rk4)
sub.plot(sol[:,0], sol[:,1])
sub.set_xlim([-1,1])
sub.set_ylim([-1,1])
sub.set_aspect('equal')

alpha = 4
sub = fig.add_subplot(2,2,4)
sub.set_title(f"$\\alpha = {alpha}$")
x0 = np.array([2,0,0,1])
t = np.linspace(0,95,10000)
sol = integrate(f, x0, t, rk4)
sub.plot(sol[:,0], sol[:,1])
sub.set_xlim([-2.5,2.5])
sub.set_ylim([-2.5,2.5])
sub.set_aspect('equal')
```

Here, the orbits are of more varied forms and are not necessarily periodic although bounded. This is an illustration of Bertrand's theorem which states that among all central potentials, only the gravitational potential ($\alpha = -1$) and the harmonic potential ($\alpha = 2$) have the property that all bounded orbits are closed.

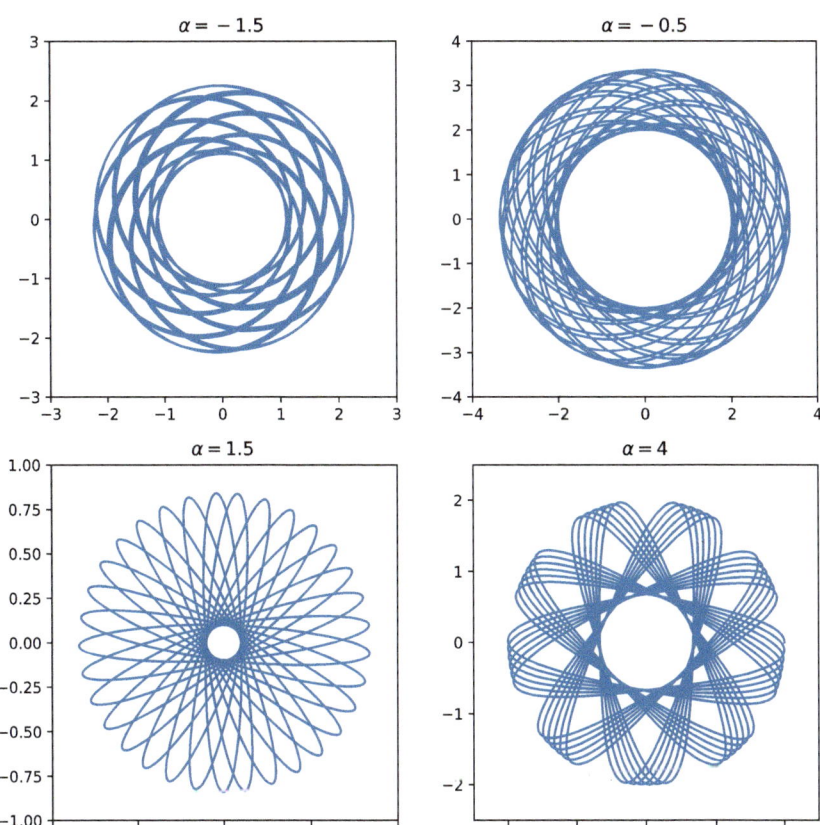

Figure 11.7    For non-physical potentials given by values of $\alpha$ different from $-1$ or 2, the orbits are no longer closed and form rosettes.

## SOLUTION 11.4   LORENZ ATTRACTOR

**a.** We need to define the function $f$:

```python
def f(t, vec, rho=28, sigma=10, beta=8/3):
    x = vec[0]
    y = vec[1]
    z = vec[2]
    return np.array([sigma*(y-x), x*(rho-z)-y, x*y-beta*z])
```

**b.** In order to be able to use the previous functions, you have to pass the parameters of the f function manually:

```python
def plot_lorenz(rho=28, sigma=10, beta=8/3):
    t = np.linspace(0,20,20*1000)
    x0 = np.array([1,1,1])
    sol = integrate(lambda t,x: f(t,x,rho,sigma,beta), x0, t,
      ↳  rk4)
    plt.title(f"$\\rho={rho}$, $\\sigma={sigma}$, and
      ↳  $\\beta={beta:.4f}$")
    plt.xlabel("$x$")
    plt.ylabel("$z$")
    plt.plot(sol[:, 0], sol[:,2])
```

This allows to draw the four requested trajectories represented in Figure 11.8:

```python
fig = plt.figure(figsize=(12,10))
for i,rho in enumerate([10,15,20,25]):
    sub = fig.add_subplot(2,2,i+1)
    plot_lorenz(rho=rho, sigma=10, beta=8/3)
```

**c.** It is necessary to begin to load SymPy:

```python
import sympy as sp
sp.init_printing()
```

then define the necessary symbols and solve the equation $f(t, \boldsymbol{x}) = \boldsymbol{0}$:

```python
rho = sp.Symbol(r"\rho")
sigma = sp.Symbol(r"\sigma")
beta = sp.Symbol(r"\beta")
x,y,z,t = sp.symbols("x y z t")
vec = sp.Matrix([x,y,z])
steady = sp.solve(f(t, vec, rho=rho, sigma=sigma, beta=beta),
  ↳  vec)
steady
```

The conclusion is that for $\rho > 1$, there are three fixed points. The three fixed points appear through a bifurcation represented in Figure 11.9:

```
plot = None
for s in steady:
    cp = sp.plot(s[0].subs(beta,8/3),(rho,0,10), show=False,
    ↪   xlabel=r"$\rho$", ylabel="$x$")
    if plot:
        plot.extend(cp)
    else:
        plot = cp
plot.show()
```

## SOLUTION 11.5   CUBIC WAVE EQUATION (!!)

**a.** The equation is equivalent to the following system:

$$\frac{\partial u}{\partial t} = v, \qquad\qquad u(0, \cdot) = u_0,$$

$$\frac{\partial v}{\partial t} = \frac{\partial^2 u}{\partial x^2} - u^3, \qquad\qquad v(0, \cdot) = v_0.$$

**b.** This is a direct application of Exercise 9.4.

**c.** The function $f$ is defined by:

$$f(t, u, v) = \begin{pmatrix} v_0 \\ v_1 \\ \vdots \\ v_{N-1} \\ v_N \\ 0 \\ \dfrac{u_0 - 2u_1 + u_2}{h^2} - u_1^3 \\ \dfrac{u_1 - 2u_2 + u_3}{h^2} - u_2^3 \\ \vdots \\ \dfrac{u_{N-2} - 2u_{N-1} + u_{N+1}}{h^2} - u_{N-1}^3 \\ 0 \end{pmatrix}$$

and therefore:

```
def f(t,uv):
    # number of discretizations
    N = len(uv)//2 -1
    # time derivative duv = {\dot{u}, \dot{v}}
    duv = np.zeros_like(uv)
    # derivative of u
    duv[:N+1] = uv[N+1:]
    # derivative of v
```

```
        duv[N+1] = 0
        duv[N+2:-1] = (uv[2:N+1] -2*uv[1:N] + uv[0:N-1])/h**2 -
         ↳  uv[1:N]**3
        duv[-1] = 0
        return duv
```

**d.** Using the Euler method, the time step must be strictly smaller than the space step to respect the Courant-Friedrichs-Lewy condition and guarantee the stability of the scheme. Here, we choose the Runge-Kutta method of order four with a time step that represents 90 % of the space step:

```
N = 1000 ; L = 100
x = np.linspace(-L,L,N+1)
h = x[1]-x[0]
t = np.arange(0,L,0.9*h)
# initial data
uv0 = np.zeros(2*N+2)
uv0[0:N+1] = np.exp(-x**2)
# resolution
sol = integrate(f,uv0,t,rk4)
```

**e.** The animation module of Matplotlib allows to create a video, capturing each frame in a loop:

```
import matplotlib.animation as manimation
# create empty figure
fig,ax = plt.subplots(figsize=(8,5))
plot, = ax.plot([], [])
ax.set_xlim(-L, L)
ax.set_ylim(-1, 1)
# write the movie at 200dpi and 15fps
writer = manimation.FFMpegWriter(fps=15)
with writer.saving(fig, "nonlinear-wave.mp4", 200):
    for i in range(len(t)):
        ax.set_title(r"Solution of the non-linear wave
         ↳  equation $t = {:.2f}$".format(t[i]))
        plot.set_data(x, sol[i][0:N+1])
        writer.grab_frame()
```

The hundredth image is represented in Figure 11.10.

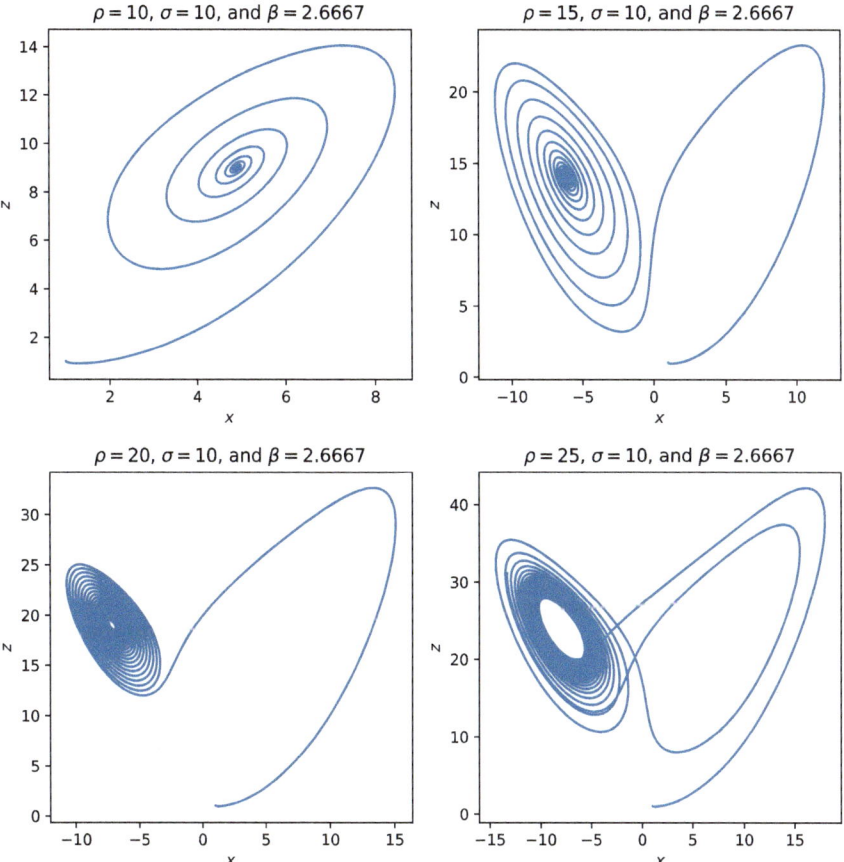

Figure 11.8    Trajectories of the Lorenz model from the initial data $(x_0, y_0, z_0) = (1, 1, 1)$ for various choices of parameters. For $\rho = 10$, the trajectory seems to converge to a point. For $\rho = 15, 20$, the trajectory seems to converge to another point. For $\rho = 25$, the trajectory oscillates between two regions.

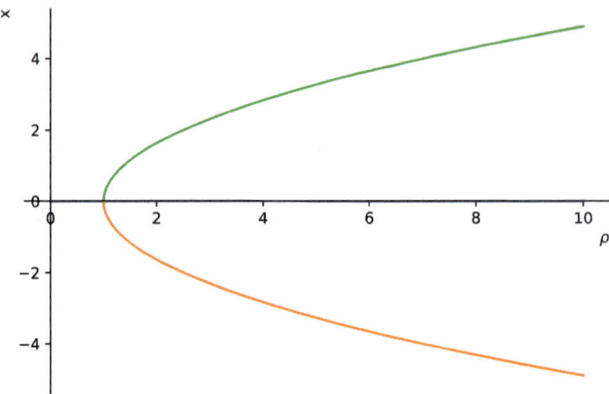

Figure 11.9    Bifurcation diagram as a function of $\rho$. When $\rho > 1$, there are two additional fixed points and the solution can converge to one or the other or oscillate around both in a chaotic way.

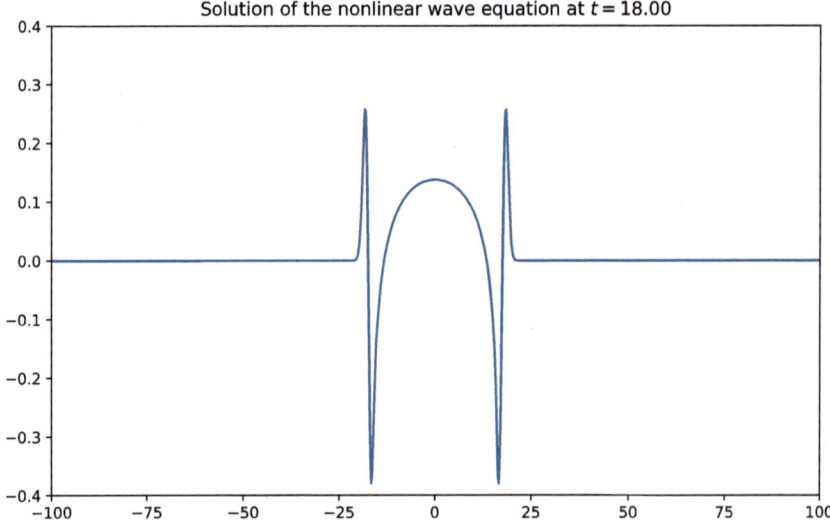

Figure 11.10    Solution of the cubic wave equation at $t = 18$ for the initial data $u_0(x) = e^{-x^2}$ and $v_0(x) = 0$.

# CHAPTER 12

# Data Science

The methodology of data science is to use available data (usally a large amount) to answers questions. Dealing with large number of data is not very practical with Python's default data structures or NumPy. First, the Pandas module specially designed for data analysis will be presented. The Pandas documentation is available at the address: `https://pandas.pydata.org/docs/`.

To load the Pandas module, it is usual to proceed as follows:

```python
import pandas as pd
```

Next, real data will be analyzed by making statistics on the proportion of numbers beginning with a certain digit, as well as determining trends. Finally, three aspects of machine learning will be examined, with handwritten digit recognition, automatic differentiation, and the use of a neural network.

## Concepts covered

- data import and analysis

- use of Pandas

- Benford's law

- least-squares methods

- image classification

- automatic differentiation

- neural networks

# EXERCISES

## EXERCISE 12.1   INTRODUCTION TO PANDAS

**Creation:** In Pandas, a data table is a `DataFrame` and constructing a table can be done manually from a dictionary, for example:

```
df = pd.DataFrame(
    {
        "Name": [
            "Braund, Mr. Owen Harris",
            "Allen, Mr. William Henry",
            "Bonnell, Miss. Elizabeth",
        ],
        "Age": [22, 35, 58],
        "Sex": ["male", "male", "female"],
    }
)
```

In Jupyter Lab, it suffices to execute the cell:

```
df
```

to display the table. One can see here, it consists of three columns (with given labels) and three rows that are labeled by integers by default.

**Columns extraction:** One single column of a `DataFrame` is called a `Series` and can be extracted easily:

```
df["Age"]
```

On such a column, statistics can be done in a simple way, for exemple, to determine the oldest people and the mean age:

```
df["Age"].max(), df["Age"].mean()
```

Two columns can also be extracted:

```
df[["Age", "Sex"]]
```

**New column creation:** One can also add additional data, for example, a new column with 100 over the ages:

```
df["Age inv"] = 100/df["Age"]
df
```

We note that such operation are elementwise so there is no need for a loop, in the same spirt as NumPy, even for more complex logic:

```
def myfunction(l):
    if l["Sex"] == "male":
        return l["Age"]+5
    else:
        return l["Age"]+2
df["Age new"] = df.apply(myfunction, axis=1)
```

**Rows extraction:** Selecting rows satisfying some criterion is very important and quite simple, for example, people being more than 30 years old:

```
df[df["Age"]>30]
```

or female being more than 30 years old:

```
df[(df["Age"]>30) & (df["Sex"]=="female")]
```

This concept is very similar to NumPy indexing. For a more complex selection, we could use the following syntax:

```
value=3
df.query('`Age inv`>@value and Sex=="male"')
```

**Slicing:** Similar to NumPy, it is possible to use slicing to select part ot the table:

```
df.loc[0:1, "Sex":"Age new"]
```

Note that the end points are included, unlike the standard Python method. However, here is slicing without endpoints:

```
df.iloc[0:2,3:5]
```

**Plotting:** Finally, data can be represented graphically in various ways. The simplest way is to do:

```
df.plot(x="Name", y=["Age","Age inv"])
```

**a.** Download the World Bank's education data in CSV format at the following address: https://data.worldbank.org/topic/education.

**b.** By inspecting the previous files, extract the data required to create a two-column table, the first containing country codes, the second country names. *Hint: Use the* read_csv *function to read a CSV file with Pandas.*

**c.** Looking at the documentation of the rename and set_index functions, rename the column labels to code and country, then set the country code as the row index. The aim is that the name of the country can then be determined from its code with the function `loc["FRA"]`.

**d.** Determine the code for Zimbabwe from the table above.

**e.** Determine the code associated with the proportion of unemployed and the proportion of people with at least a master's degree from the data downloaded above.
*Hint: Search for the "Unemployment" and "Master" strings in the file* `Metadata_|Indicator....`

**f.** Plot the evolution of the unemployment rate in France.
*Hint: The file* `API...` *contains a header and really only starts at line 5; use the* `skiprows=4` *option to ignore the header.*

**g.** Write a function `time_plot(code, countries)` to plot the time evolution of the code indicator for the countries in the `countries` list. Test it on the unemployment rate and the proportion of people with at least a master's degree for different countries.

## EXERCISE 12.2   BENFORD'S LAW

Benford's law predicts that statistically in a list of given numbers, the probability that one of these numbers begins with the digit 1 is greater than the probability that it begins with the digit 9. More precisely, Benford's law predicts that the probability that a number begins with the digit $d$ is:

$$p_d = \log_{10}\left(1 + \frac{1}{d}\right),$$

where $\log_{10}$ is the logarithm in base 10. It is possible to verify that Benford's law is the only one that remains invariant by change of units, *i.e.*, by multiplying the numbers of the list by a constant, the previous probabilities remain unchanged.

**a.** Write a function `firstdigit(n)` which for a given number n returns its first digit and a function `occurrences(lst)` which returns the number of occurrences of the first digits in `lst`.
*Hint: Make the* `occurrences` *function work even if the list contains zeros by ignoring them.*

**b.** Check if Benford's law seems to be satisfied for the sequence of numbers $(2^n)_{n\in\mathbb{N}}$ by comparing the empirical histogram with Benford's law.

**c.** Check if Benford's law seems to be satisfied for the sequence of numbers $(3n + 1)_{n\in\mathbb{N}}$.

**d.** By going to the INSEE website at the address: `https://www.insee.fr/fr/statistiques/7631680`, download the file in CSV format containing the French population data by sex and age grouped (POP1A). Import this data to have the population by postal code, sex, and age group.
*Hint: Documentation on how to read files is available at the address:* `https://docs.python.org/3/tutorial/inputoutput.html#reading-and-writing-files`.

**e.** Determine if the list of all populations by postal code, sex, and age follows Benford's law.

**f.** Sum the previous data to obtain the list of populations by postal code and determine if it follows Benford's law.

**g.** Repeat the previous two questions, but using the POP1B population file with ungrouped ages and using Pandas.

**h.** ‼ By going to the INSEE website or elsewhere download your favorite dataset and test if it follows Benford's law.
*Hint: Use, for example, the detailed French government accounts available at the address:* https://www.data.gouv.fr/fr/datasets/donnees-de-comptabi lite-generale-de-letat/.

## EXERCISE 12.3   LEAST SQUARES METHOD

**a.** Reuse the INSEE data on the French population by sex and age POP1B. Write a function pop(code) that returns, as a list or NumPy vector, the population by age, without distinction between men and women, in the municipality with postal code code. Use this function to determine the total number of people living in postal code 75102.

**b.** Write a function plot_pop(code) to plot the population fractions (normalized by the total population of the municipality) as a function of age, without distinction between men and women, in the municipality with postal code code. Test for municipalities with postal codes 13201 and 75102.
For a given municipality, if we denote by $p(a)$ the fraction of the population of age $a$, we look for the coefficients $r_0, r_1, r_2$ such that the law:

$$p(a) = r_0 + r_1 a + r_2 a^2$$

is best satisfied for ages $a \geq 25$. To do this, we solve the least squares problem:

$$\min_{r \in \mathbb{R}^3} \|Xr - p\|^2,$$

where by noting the vector of ages $a = (25, 26, \dots, 100)$, $X$ is the matrix of size $76 \times 3$ such that $X_{i,1} = 1$, $X_{i,2} = a_i$, $X_{i,3} = a_i^2$, and $p$ is the vector of populations by age $p_i = p(a_i)$. The solution to this problem is $r = (r_0, r_1, r_2) \in \mathbb{R}^3$. This solution satisfies the equation:

$$X^\mathsf{T} X r = X^\mathsf{T} p,$$

where $X^\mathsf{T}$ is the transpose of $X$.

**c.** For the municipality with postal code 13201, form the matrix $X$ and the vector $p$, then determine the solution $r$.

**d.** For municipalities with postal codes 13201 and 75102, plot the theoretical curve $r_0 + r_1 a + r_2 a^2$ as a function of age over the data.

## EXERCISE 12.4   HANDWRITTEN NUMBER RECOGNITION

The aim of this exercise is to classify images of handwritten numbers, *i.e.*, to recognize handwritten numbers. This is a simple example of machine learning and one of the first industrial applications for automatic reading of cheques or postal codes.

The scanned handwritten digits dataset comes from the UCI ML repository. There are two ways to import this dataset. If the scikit-learn package (import name sklearn) is installed, simply run:

```
from sklearn.datasets import load_digits
digits = load_digits()
X, y = digits.data, digits.target
```

If the sklearn module is unavailable, the following commands can be used instead to load data:

```
import urllib.request, gzip, io
# download link
url = "https://raw.githubusercontent.com/scikit-learn/
    scikit-learn/main/sklearn/datasets/data/digits.csv.gz"
# download gz file
file = urllib.request.urlopen(url)
# extracts the gz file
file = gzip.GzipFile(fileobj=io.BytesIO(file.read()))
# import txt file
digits = np.loadtxt(file, delimiter =',')
# extract images and labels
X, y = digits[:,:-1], digits[:,-1]
```

In both cases, X is a NumPy array containing numerous examples of handwritten digitized digits in an $8 \times 8$ pixel image stored as an array of 64 integers stored as floats. The y variable contains the integer between 0 and 9 corresponding to the digitized digit. This is referred to as *label*.

**a.** Determine the dimensions of X and y and deduce the number of examples contained in the database.

**b.** Display the data contained in X associated with the index idx=12. This is the 12th line of the X table, starting the numbering at zero.

**c.** Using NumPy function reshape and Matplotlib function imshow, display the image with index idx=12. You can use the cmap='gray' argument in the imshow call to display the result in grayscale. Which digit is encoded in this way?
For each digit class (from 0 to 9), the idea is to calculate its centroid, *i.e.*, the "average" representation of a class.

**d.** Define the X and y sub-tables corresponding to all digitized 0 digits.

**e.** For all the zeros from the previous question, calculate the mean value for each pixel to define the "average zero".

**f.** For all the digits from 0 to 9, plot the associated average image on the same line as shown in Figure 12.1.
*Hint: Use Matplotlib's* subplot *function.*

mean 0  mean 1  mean 2  mean 3  mean 4  mean 5  mean 6  mean 7  mean 8  mean 9

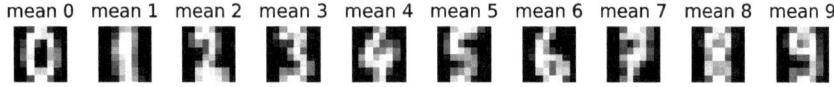

Figure 12.1    Averages of digitized figures.

Finally, we will implement our own classifier: for a new digitized digit image, we will predict the class whose average digit is closest. To do this, we divide our dataset into two parts of similar size: the first part will serve as training data (X_ train and y_train) and the second part will serve as test data (X_test and y_ test).

**g.** Define variables: X_train, y_train, X_test and y_test.

**h.** For each digit in the training set, calculate the centroids (*i.e.*, the average digits) of the classes from 0 to 9. Note the variable containing the set of averages centroids_train.

**i.** Write a function which, given a number in the test set (X_test), returns the centroid of the nearest centroids_train in the Euclidean norm.

**j.** Finally, evaluate whether the digit thus obtained corresponds to the true digit using y_test and deduce an estimate of the percentage of correct predictions on the test set.

## EXERCISE 12.5   AUTOMATIC DIFFERENTIATION (!)

The aim of this exercise is to introduce one of the fundamental building blocks of machine learning: automatic differentiation. This is a technique for calculating derivatives or gradients of Python functions in a way that is virtually transparent to the user. Given a certain Python function f(x), the aim of automatic differentiation is to make it virtually as easy for the user to evaluate the derivative of f at a point x=1 as it is to do f(1), even if the function $f$ is complicated. This is the fundamental building block enabling machine learning to learn parameters by optimizing complicated nonlinear functions.

The idea behind automatic differentiation is to use the derivation rule for compound functions and knowledge of the derivatives of basic functions. In fact, a Python function is "just" a composition of basic functions (or instructions).

For sake of simplicity, we are only interested here in the composition of functions on $\mathbb{R}$. For any $i \in \{0, 1, \dots, n\}$, let $f_i : \mathbb{R}$ be an elementary function whose derivative is known. Consider the composition of the first $i \leq n$ functions:

$$F_i = f_i \circ f_{i-1} \circ \cdots \circ f_1 \circ f_0.$$

The derivative is given by the composition rule (or chain rule):

$$F_i' = (f_i \circ F_{i-1})' = (f_i' \circ F_{i-1}) \cdot F_{i-1}'.$$

This gives a recursive way of calculating $F_n'$ with the anchor $F_0' = f_0'$. To calculate the value of $F_n(x)$ for a given $x$, Pyhon will intrinsically calculate successively $F_0(x)$, then $F_1(x)$, $F_2(x)$, up to $F_n(x)$. Automatic differentiation consists of evaluating $F_0'(x)$, $F_1'(x)$ up to $F_n'(x)$ at the same time or later.

Note that automatic differentiation is neither a numerical approximation nor symbolic calculation. In fact, the value of the derivative is determined exactly (to machine precision), which is not the case with numerical approximation:

$$f'(x) \approx \frac{f(x+h) - f(x)}{h},$$

with some $h > 0$ small. It is not symbolic calculation either, as there are no symbols in automatic differentiation, only real numbers. Automatic differentiation calculates the value of the derivative at a given point, whereas symbolic differentiation does this for any symbol.

**a.** The first step is to define the derivatives of the elementary functions. To do this, build the tuples $\tilde{f} = (f, f')$ manually for the elementary functions sin, cos, op : $x \mapsto -x$, inv : $x \mapsto x^{-1}$ et square : $x \mapsto x^2$.

**b.** A composition of functions $F_n = f_n \circ f_{n-1} \circ \cdots \circ f_1 \circ f_0$ will be stored in Python as the list of tuples $[\tilde{f}_0, \tilde{f}_1, \dots, \tilde{f}_n]$. In Python, define the composition corresponding to the function:

$$F(x) = \cos\left(\frac{1}{\sin(-\cos x^2)^2}\right).$$

**c.** Write a function `eval(list, x)` which, given a list of tuples defining a composition of functions $F_n$, returns $F_n(x)$. Test on the previous example.

**d.** Write a function `autodiff(list, x)` which, given a tuple list defining a composition of functions $F_n$, returns $F_n(x)$ and $F_n'(x)$. Test again on the same example.

The previous approach only allows the composition of functions of one variable, which is very limiting, as sum and multiplication are functions of two variables. The idea is to be able to consider more complicated functions as well, for example:

$$G(x) = \frac{\cos(x)}{\sin(x) + \cos(x)\sin(x^2)}.$$

**e.** As before, implement the sum function add : $x, y \mapsto x + y$ and the multiplication function mult : $x, y \mapsto xy$ in tuple form.

**f.** The previous function $G$ can no longer be represented in Python as a list of compositions, as it has the structure of a graph:

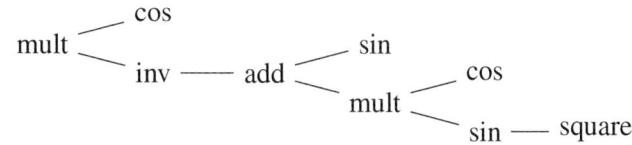

We choose to store it in Python as a list of lists, with the first element of each list being the function to be applied, and the arguments being the following children. For the example $G$ above:

```
myG = [mult, [cos], [inv, [add, [sin], [mult, [cos], [sin,
    [square]]]]]]
```

Write the previous simple composition $F$ in this new form, as well as the function:

$$H(x) = \frac{\cos(x^2) + \sin x}{\cos x + 2 \sin(x^{-1})}.$$

**g.** Write a function `eval(list,x)` which evaluate an expression in the form of a list of lists described above at **x**. Test on functions $F$, $G$, and $H$.

**h.** !! Write a function `autodiff(list,x)` which, in addition to returning the evaluation of the function at **x**, also returns the evaluation of its derivative at **x**.

**i.** !! The previous implementation is not very usable concretely. In practice, automatic differentiation is coded by overloading basic operations in order to be relatively transparent to the user. Using the package JAX or PyTorch, determine the derivative of the following function by automatic differentiation at **x=0.4**:

```
def f(x):
    for i in range(50):
        if x>0.5:
            x = 3.7*x*(1-x)
        else:
            x = 3*x*(1-x)
    return x
```

Finally, compare with numerical differentiation.
*Hint: The documentation on automatic differentiation with JAX is at the address:* `https://jax.readthedocs.io/en/latest/automatic-differentiation.html` *and with PyTorch at the address:* `https://pytorch.org/tutorials/beginner/blitz/autograd_tutorial.html`.

## EXERCISE 12.6   NEURAL NETWORK (!)

The aim of this exercise is to introduce the concept of neural network for finding a real function for which only its noisy evaluation is known. More precisely, consider the function $f : [0,1] \to \mathbb{R}$ defined by:

$$f(x) = 1 + \sin(4 \cos x)^2,$$

and consider the data generated by 500 noisy evaluations of $f$:

```
rng = np.random.default_rng(123456)
f = lambda x: 0.1 + np.sin(4*np.cos(x))**2
x = rng.random(500)
y = f(x) + rng.normal(0,0.1,500)
```

The aim is to forget that the data were generated in this way and to find a function approximating $f$ from $x$ and $y$ alone. To achieve this, a single-layer neural network will be used, and we speak of learning the function from the data. The principle is to construct the learned function in the form:

$$f_\omega(x) = \frac{1}{n} \sum_{i=0}^{n-1} w_i \sigma(a_i x + b_i),$$

where $\omega = (a, b, w) \in \mathbb{R}^n \times \mathbb{R}^n \times \mathbb{R}^n$ are parameters to be determined and $\sigma$ the sigmoid function:

$$\sigma(x) = \frac{1}{1 + e^{-x}}.$$

This is a single-layer neural network with $n$ neurons. The principle is to sum $n$ nonlinear functions (the sigmoids) with input weights $a = (a_i)_{i=0}^{n-1}$, biases $b = (b_i)_{i=0}^{n-1}$, and output weights $w = (w_i)_{i=0}^{n-1}$. Parameters are chosen to minimize the following cost function:

$$J(\omega) = \sum_{k=0}^{499} \left(f_\omega(x_k) - y_k\right)^2.$$

The strategy for minimizing $J$ on the parameters $\omega$ is to perform a gradient descent starting with random values $\omega_0$ of the parameters and then successively defining:

$$\omega_{i+1} = \omega_i - \eta J'(\omega_i),$$

where $J'(\omega)$ is the gradient of $J(\omega)$ with respect to the parameters and $\eta \in (0, 1]$ is a parameter called the learning rate. The idea of the gradient descent algorithm is to move the parameters in the direction of greatest gradient in order to minimize $J(\omega)$. This is known as parameters learning.

**a.** Plot the data $x$ and $y$ and the function $f$.

**b.** Determine the value of the cost function $J$ for the function $f$:

$$J_f = \sum_{k=0}^{499} \left(f(x_k) - y_k\right)^2.$$

**c.** Define the sigmoid function $\sigma$ in Python and its derivative $\sigma'$, and represent the sigmoid graphically.

**d.** Implement in Python a function F(x,omega) corresponding to the function $f_\omega(x)$. Make the function F(x,omega) vectorized, *i.e.*, if $x = (x_0, x_1, ..., x_{k-1}) \in \mathbb{R}^k$, then the function should return $(f_\omega(x_0), f_\omega(x_1), ..., f_\omega(x_{k-1}))$.

**e.** Calculate by hand the gradient of $f_\omega(x)$ with respect to $\omega$ (and not with respect to $x$) and implement this gradient in Python, taking care that it is also vectorized.

**f.** Implement in Python $J(\omega)$ and its gradient $J'(\omega)$.

**g.** With learning rate $\eta = 0.01$ and four neurons, learn the parameters $\omega \in \mathbb{R}^{12}$ that tend to minimize $J$. Compare the value of the cost function $J(\omega)$ of the learned function $f_\omega$ with the value of the cost function $J_f$ of the function $f$. Plot the function $f_\omega$ as a function of $x$ for these parameters and compare with the function $f$.

# SOLUTIONS

## SOLUTION 12.1   INTRODUCTION TO PANDAS

**a.** You can download the file using your computer's browser, then decompress it, but it's also possible to do this with Python:

```python
import urllib.request, zipfile, io
url = "https://api.worldbank.org/v2/en/topic/|
  ↳  4?downloadformat=csv"
# download zip file
file = urllib.request.urlopen(url)
# extract the zip file
# (io.ByteIO creates a pseudo file for in-memory
  ↳  decompression)
zipper = zipfile.ZipFile(io.BytesIO(file.read()))
# lists files present
zipper.namelist()
```

**b.** The file `Metadata_Country`... contains country information. Simply extract the file and import it:

```python
# opens csv file with countries
filename = [l for l in zipper.namelist() if
  ↳  l.startswith("Metadata_Country")][0]
f = zipper.open(filename)
meta_country = pd.read_csv(f)
meta_country
```

and finally extract the requested data:

```python
countries = meta_country[["Country Code", "TableName"]]
```

**c.** First, rename the columns, then transform the first column into an index:

```python
countries = countries.rename(columns={"Country Code": "code",
  ↳  "TableName": "country"})
countries = countries.set_index("code")
countries
```

This is then used to determine the full name of the country from its code:

```python
countries.loc["FRA"]
```

**d.** First, select the lines containing Zimbabwe, then take the first line and extract the code:

```
countries[countries["country"] == "Zimbabwe"].iloc[0].name
```

**e.** The first step is to import the file Metadata_Indicator...:

```
filename = [l for l in zipper.namelist() if
 ↳  l.startswith("Metadata_Indicator")][0]
meta = pd.read_csv(zipper.open(filename))
meta
```

then search for "Unemployment" in the column INDICATOR_NAME:

```
meta[meta['INDICATOR_NAME'].str.contains('Unemployment')]
```

The code associated with unemployment is therefore SL.UEM.TOTL.ZS. The same applies to the proportion of people with at least a master's degree:

```
meta[meta['INDICATOR_NAME'].str.contains('Master')]
```

and we find SE.TER.CUAT.MS.ZS.

**f.** The first step is to import the data, ignoring the first header lines:

```
filename = [l for l in zipper.namelist() if
 ↳  l.startswith("API")][0]
data = pd.read_csv(zipper.open(filename), skiprows=4)
data
```

Then, select unemployment data for France:

```
df = data.query('`Country Code` == "FRA" and `Indicator Code`
 ↳  == "SL.UEM.TOTL.ZS"')
df
```

The problem now is that the data is in one row, whereas we would like to have the years as a column. To do this, select the range of years to be extracted and take the transpose:

```
df = df.loc[:,"1990":"2022"].T
df
```

Finally, graphical representation is easy:

```
df.columns = ["France"]
df.plot()
```

The first instruction simply renames the column name to "France" and thus displays the correct label on the graph.

**g.** The following function can be used to graphically represent the temporal evolution of an indicator for several countries:

```python
def time_plot(code, countries, title=""):
    df = data.query('`Country Code` in @countries and
     ↳  `Indicator Code` == @code')
    df = df.set_index('Country Name')
    df = df.loc[:,"1990":"2022"].T
    df.plot(title=title)
```

Finally, to test the unemployment rate:

```python
time_plot("SL.UEM.TOTL.ZS", ["FRA", "USA", "CHE", "ZWE"],
 ↳  title="Evolution of the unemployment rate")
```

and the proportion of masters:

```python
time_plot("SE.TER.CUAT.MS.ZS", ["FRA", "USA", "CHE", "ZWE"],
 ↳  title="Evolution of the proportion of master's degrees")
```

where some data is unfortunately lacking.

## SOLUTION 12.2   BENFORD'S LAW

**a.** Just convert the number to a string, then select the first one and convert it back to an integer:

```python
def firstdigit(n):
    return int(str(n)[0])
```

To determine the number of occurrences of each number:

```python
def occurrences(lst):
    out = 9*[0]
    for d in lst:
        if d != 0:
            out[d-1] += 1
    return out
```

**b.** To generate the first digits of the first $10^4$ numbers of the sequence:

```python
N = 10_000
liste = [2**n for n in range(N)]
freq = occurrences(map(firstdigit, liste))
```

Such an approach is rather slow and hardly applicable to more terms. The reason is the time needed to convert a large integer into a decimal representation. To do statistics on more terms, one possibility is to use the decimal module which keeps a decimal representation of the numbers and easily allows to take into account $10^5$ terms or more:

```
import decimal
N = 100_000
liste = [decimal.Decimal(2)**n for n in range(N)]
freq = occurrences(map(firstdigit, liste))
```

Finally, we define a function that allows us to produce the histogram of a list of frequencies:

```
import numpy as np
import matplotlib.pyplot as plt
def compare(freq, title=""):
    N = sum(freq)
    plt.figure(figsize=(8,5))
    plt.title(title)
    # occurrences according to Benford's law
    benford = [N*np.log10(1+1/d) for d in range(1,10)]
    # bar width
    width = 0.3
    plt.bar(np.arange(1,10)-width/2, benford, width,
      ↳  label="Benford's law")
    plt.bar(np.arange(1,10)+width/2, freq, width,
      ↳  label="Data")
    plt.xticks(range(1,10))
    plt.legend()
```

This allows us to see in Figure 12.2 that Benford's law is extremely well satisfied by the sequence $2^n$:

```
compare(freq, title=r"Frequencies of $2^n$ for $0\leq n \leq
  ↳  10^5$")
```

**c.** Benford's law does not seem to be satisfied in this case as shown in Figure 12.3:

```
N = 100_000
liste = [3*n+1 for n in range(N)]
freq = occurrences(map(firstdigit, liste))
compare(freq, r"Frequencies of $3n+1$ for $0\leq n \leq
  ↳  10^5$")
```

**d.** The following implementation automatically downloads the file, decompresses it into memory, and reads it:

```
import urllib.request, zipfile, io
# file to download
url = "https://www.insee.fr/fr/statistiques/fichier/7631680/
  ↳  TD_POP1A_2020_csv.zip"
# download the zip file
file = urllib.request.urlopen(url)
# extract the zip file
```

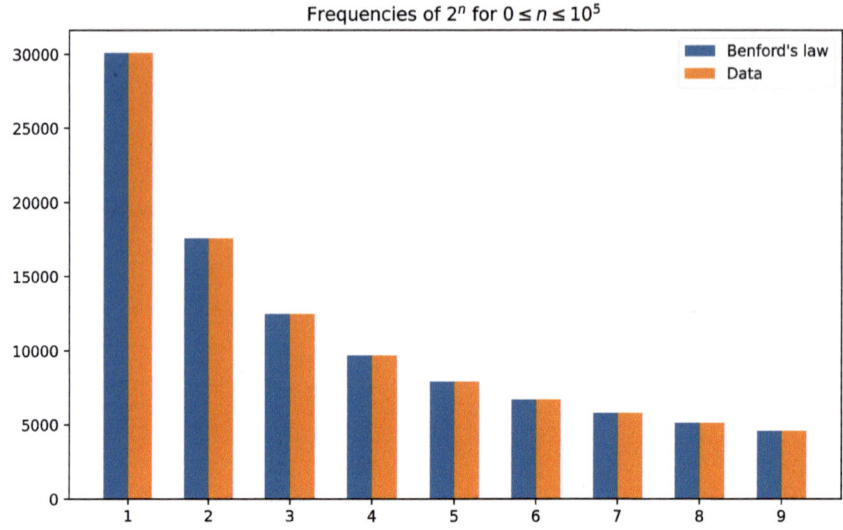

Figure 12.2    Number of occurrences of each digit in the sequence $2^n$. For this sequence, Benford's law is very well satisfied.

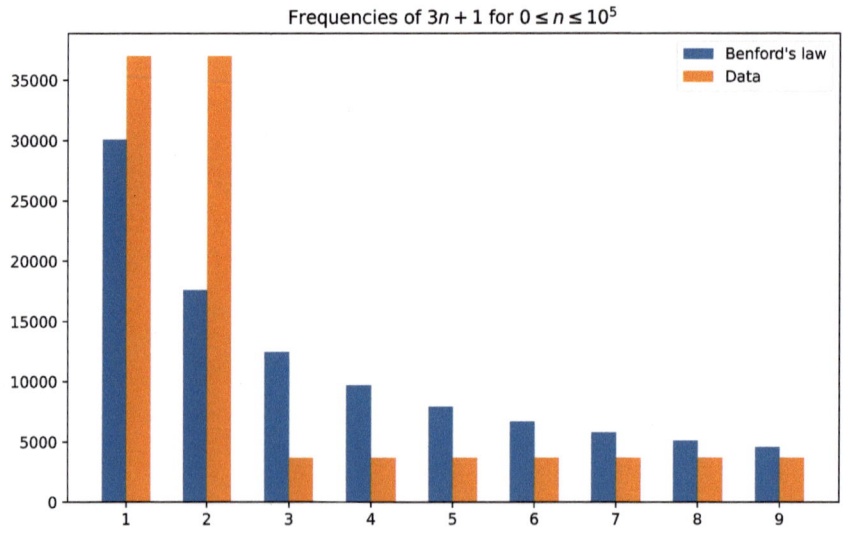

Figure 12.3    Benford's law is not satisfied on the first numbers of the sequence $3n + 1$.

```
# (io.ByteIO create a virtual file to uncompress in memory)
zipper = zipfile.ZipFile(io.BytesIO(file.read()))
# open the CSV file
f = zipper.open("TD_POP1A_2020.csv")
```

In an equivalent way, it is possible to place the downloaded and decompressed file by hand in the current directory and use it instead:

```
f = open("TD_POP1A_2020.csv", "rb")
```

Then, you have to make a loop to import the data into a list:

```
# list to store the results (postal code, age, population)
data = []
# read the first line with labels
f.readline()
# loop on all remaining lines
for line in f:
    # decode the line in UTF-8 and split the fields
    line = line.decode().split(";")
    # consider only valid lines
    if len(line)==6:
        code = line[1] # postal code
        sex = int(line[3]) # sex
        age = int(line[4]) # age
        pop = float(line[5]) # population
        # add to the list of data
        data.append((code,sex,age,pop))
```

**e.** The data follow Benford's law rather well as shown in Figure 12.4:

```
lst = [firstdigit(v[-1]) for v in data]
freq = occurrences(lst)
compare(freq, "Population by postal code, sex, and age
  ↳  group")
```

**f.** When aggregated by municipality, the data still follow Benford's law well:

```
dic = {}
for v in data:
    # index the dictionary of the postal codes and sum the
      ↳  populations with same codes
    dic[v[0]] = dic.get(v[0],0) + v[-1]
lst = [firstdigit(p) for p in dic.values()]
freq = occurrences(lst)
compare(freq, "Population by postal code")
```

**g.** Using Pandas, the import is immediate:

```python
import pandas as pd
# file to download
url = "https://www.insee.fr/fr/statistiques/fichier/7631680/
    ↪  TD_POP1B_2020_csv.zip"
# pandas download and uncompress the file automatically
data = pd.read_csv(url, sep=";", dtype={"CODGEO":str})
```

The population by ungrouped age follows Benford's law less well as represented in Figure 12.5 because small numbers are overrepresented in small municipalities:

```python
freq = occurrences(data["NB"].apply(firstdigit))
compare(freq, "Population by the postal code, sex, and age")
```

With Pandas, the aggregation of data by municipality is simplified:

```python
s = data.groupby(["CODGEO"])["NB"].sum()
freq = occurrences(s.apply(firstdigit))
compare(freq, "Population by postal code")
```

**h.** For example, for detailed French government accounts:

```python
url = "https://www.data.gouv.fr/fr/datasets/r/f2b03519-de6f-
    ↪  4ca3-9eee-6a6004a414c9"
f = urllib.request.urlopen(url)
data = []
# read the first line with labels
f.readline()
for line in f:
    # decode the line in ISO-8859-1 and split the fields
    line = line.decode("iso-8859-1").split(";")
    # 2022 accounts
    val = line[8]
    # remove punctuation and minus signs
    for c in [",", "-", " "]:
        val = val.replace(c,"")
    # do not keep missing values
    if val != "":
        data.append(int(val))
```

Figure 12.6 shows that the 2022 accounts follow Benford's Law:

```python
liste = [firstdigit(v) for v in data]
freq = occurrences(liste)
compare(freq, "Accounts 2022 of the French government")
```

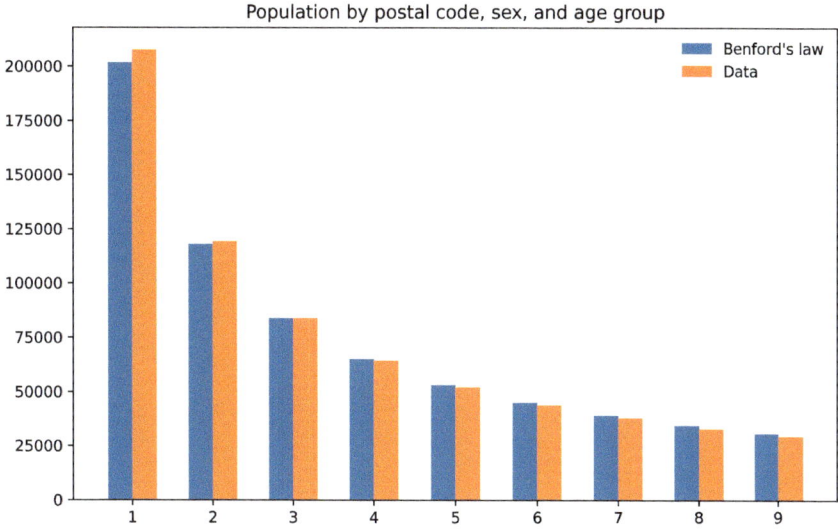

Figure 12.4   The French population in each municipality and for each age group follows Benford's law.

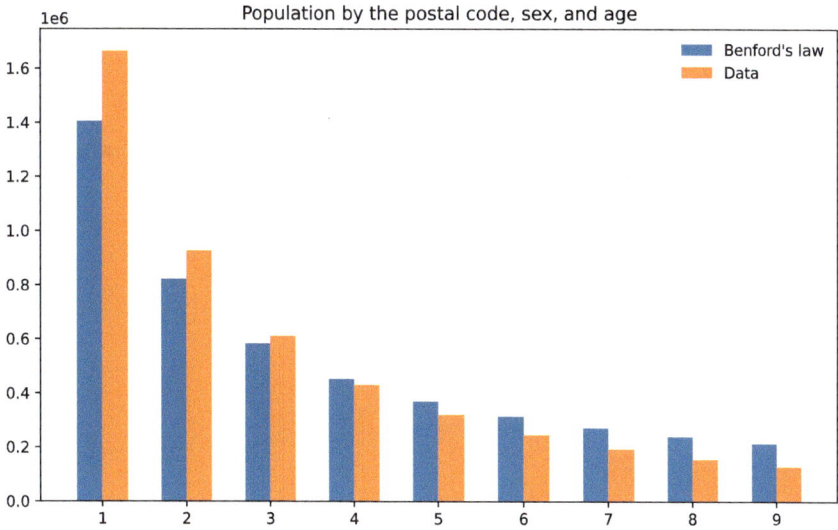

Figure 12.5   By taking the list of populations by municipality and by year of birth, Benford's law is less well satisfied. Indeed, there is not enough aggregation of data.

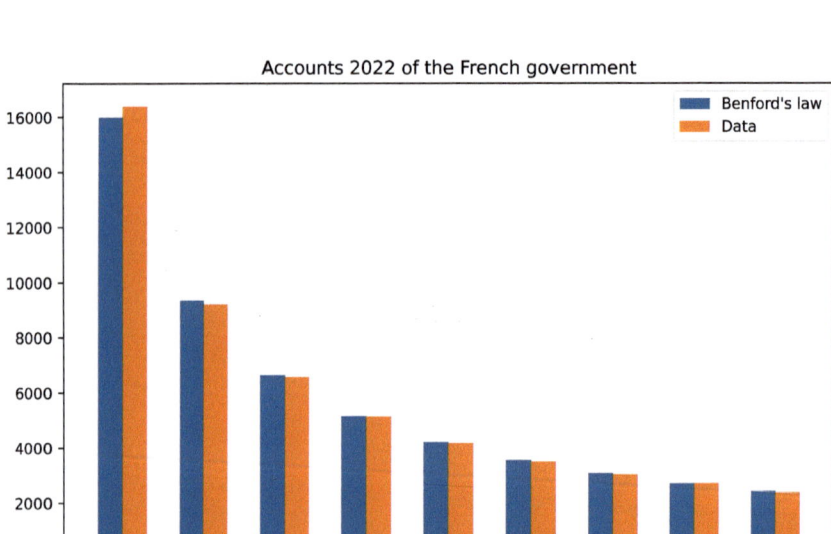

Figure 12.6 The general accounts of France follow Benford's law surprisingly well.

## SOLUTION 12.3   LEAST SQUARES METHOD

**a.** As before, with Pandas import is immediate:

```
url = "https://www.insee.fr/fr/statistiques/fichier/7631680/|
    ↳  TD_POP1B_2020_csv.zip"
data = pd.read_csv(url, sep=";", dtype={"CODGEO":str})
```

Then, write the function required to group them by postcode:

```
def pop(code):
    return data[data['CODGEO'] ==
        ↳  code].groupby('AGED100')['NB'].sum().values
```

Obtaining the total population of the requested municipality is very simple:

```
pop('75102').sum()
```

**b.** This displays the previously calculated data:

```
import matplotlib.pyplot as plt
def plot_pop(code):
    pop_code = pop(code)
    plt.plot(pop_code/pop_code.sum(),'.',label=code)
    plt.legend()
plot_pop('13201')
plot_pop('75102')
```

**c.** Using NumPy to extract the requested age range:

```
agemin = 25
a = np.arange(agemin,101)
pop_code = pop('13201')
frac_code = pop_code/pop_code.sum()
p = frac_code[agemin:]
X = np.array([np.ones(len(a)),a,a**2]).T
r = np.linalg.solve(np.dot(X.T,X),np.dot(X.T,p))
```

**d.** It is all about wrapping the previous code to get Figure 12.7:

```
def plot_reg(code):
    agemin = 25
    a = np.arange(agemin,101)
    pop_code = pop(code)
    frac_code = pop_code/pop_code.sum()
    p = frac_code[agemin:]
    X = np.array([np.ones(len(a)),a,a**2]).T
    r = np.linalg.solve(np.dot(X.T,X),np.dot(X.T,p))
    plt.plot(a, p,'.', label=f"Data {code}")
```

```
    plt.plot(a, np.dot(X,r), label=f"Regression {code}")
    plt.legend()
plt.figure(figsize=(8,5))
plt.title("Second-order regression on population density by
  ↵  municipality")
plot_reg('13201')
plot_reg('75102')
```

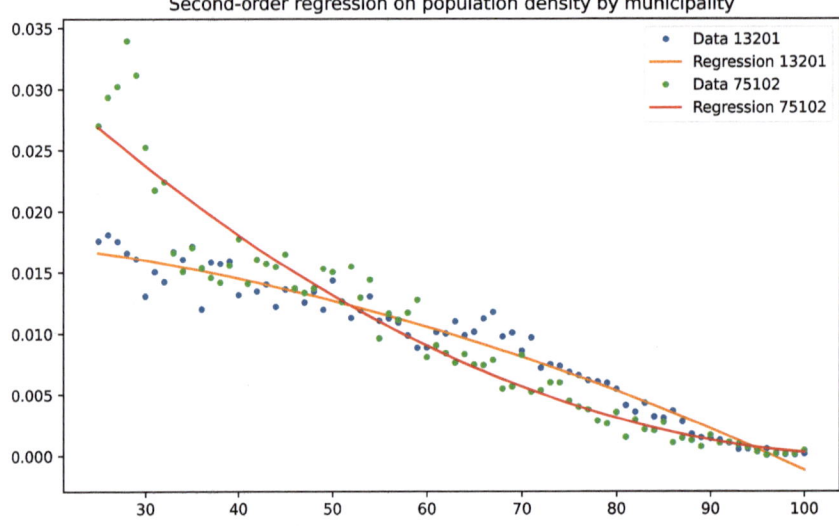

Figure 12.7   Second-order regression on population age structure for two French municipalities.

## SOLUTION 12.4   HANDWRITTEN NUMBER RECOGNITION

**a.** The `len` function can be used to determine array dimensions:

```
len(X)
len(y)
len(X[0])
```

or the shape property of NumPy arrays:

```
X.shape
y.shape
```

This means that there are 1797 examples in the database.

**b.** Simply display the 12th element of X:

```
idx=12
X[idx]
```

**c.** We need to convert the vector of size 64 into a matrix of size $8 \times 8$ for display:

```
plt.imshow(X[idx].reshape(8,8),cmap='gray')
plt.title(f'Chiffre {y[idx]}')
```

and so the coded digit is indeed a two, as indicated by `y[idx]`.

**d.** With indexing, this is trivial:

```
X0 = X[y==0]
y[y==0]
```

**e.** Use the mean function to average along only one axis to give the "average zero":

```
X0mean = np.mean(X0, axis=0)
plt.imshow(X0mean.reshape(8,8),cmap='gray')
plt.title('mean 0')
```

**f.** Iterating on numbers from 0 to 9:

```
plt.figure(figsize=(8,1))
for i in range(10):
    mean=np.mean(X[y==i],axis=0)
    plt.subplot(1,10,i+1)
    plt.imshow(mean.reshape(8,8),cmap='gray')
    plt.title(f'mean {i}')
    plt.axis('off')
plt.tight_layout()
```

**g.** We calculate the half-length, staying within the integers, and then calculate the requested variables:

```
L = len(X)//2
X_train = X[:L]
y_train = y[:L]
X_test = X[L:]
y_test = y[L:]
```

**h.** It is almost done with the above:

```
centroids_train = [np.mean(X_train[y_train==i],axis=0) for i
↳   in range(10)]
```

**i.** The following function estimates which digit a `xt` vector is closest to:

```
def estimation(xt):
    dist = np.array([np.linalg.norm(xt-c) for c in
    ↳   centroids_train])
    chiffre = np.argmin(dist)
    return chiffre
```

**j.** We define the vector with the estimates of the numbers tested:

```
y_estim = np.apply_along_axis(estimation,1,X_test)
```

which allows `y_estim` to be compared with `y_test`:

```
np.mean(y_estim==y_test)
```

to determine that the algorithm predicts correctly in almost 90% of cases.

## SOLUTION 12.5   AUTOMATIC DIFFERENTIATION (!)

**a.** The requested tuples are defined using NumPy functions:

```
sin = (np.sin, np.cos)
cos = (np.cos, lambda x: -np.sin(x))
op = (lambda x: -x, lambda x: -1)
inv = (lambda x: 1/x, lambda x: -1/x**2)
square = (lambda x: x**2, lambda x: 2*x)
```

Note that $-$np.sin is not defined in Python and that the derivative of $x \mapsto -x$ is the function $x \mapsto -1$ and not just the number $-1$.

**b.** It is just a matter of listing the sequence of operations:

```
myF = [square, cos, op, sin, square, inv, cos]
```

**c.** This involves iterating over the list of tuples and applying the function each time:

```
def eval(liste, x):
    for f,fp in liste:
        x = f(x)
    return x
```

On the previous example, this gives:

```
eval(myF,2)
```

which corresponds to the classic NumPy evaluation:

```
F = lambda x: np.cos(1/np.sin(-np.cos(x**2))**2)
F(2)
```

**d.** This involves implementing the recurrence formula $F_i' = (f_i' \circ F_{i-1}) \cdot F_{i-1}'$:

```
def autodiff(liste, x):
    Fi = x
    Fpi = 1
    for f,fp in liste:
        Fpi = fp(Fi)*Fpi
        Fi = f(Fi)
    return Fi,Fpi
```

On the same example:

```
autodiff(myF,2)
```

and a numerical approximation of the derivative gives a similar result:

```
h = 1e-10
(F(2+h)-F(2))/h
```

**e.** The idea is to implement them as functions of two variables, the second part of the tuples being the gradient:

```
add = (lambda x,y: x+y, lambda x,y: [1,1])
mult = (lambda x,y: x*y, lambda x,y: [y,x])
```

**f.** Compared with the previous representation of $F$, the order is reversed, and more parentheses are required:

```
myF = [cos, [inv, [square, [sin, [op, [cos, [square]]]]]]]
```

For the function $H$, its graph is:

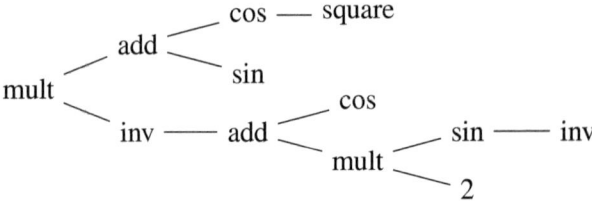

and therefore:

```
two = (lambda x: 2*x, lambda x: 2)
myH = [mult, [add, [sin], [cos, [square]]], [inv, [add,
    [cos], [mult, [two], [sin, [inv]]]]]]
```

**g.** The easiest way is to define a recursive function that evaluates the function on leaves (*i.e.*, nodes without children) and then on children:

```
def eval(liste, x):

    # function and its derivative
    f, fp = liste[0]

    # leave: evaluate f on x
    if len(liste) == 1:
        return f(x)
    # otherwise evaluate f on children
    else:
        Fi = [eval(liste[i], x) for i in range(1,
            len(liste))]
        return f(*Fi)
```

On the function $F$, we recover the previous result:

```
eval(myF,2)
```

For the function $G$, we obtain:

```
eval(myG,2)
```

which is in line with the classic assessment:

```
G = lambda x: np.cos(x) / (np.sin(x) +
    np.cos(x)*np.sin(x**2))
G(2)
```

The same applies to $H$:

```
H = lambda x: (np.sin(x)+np.cos(x**2)) / (np.cos(x) +
    2*np.sin(1/x))
eval(myH,2) - H(2)
```

**h.** The tuple $(f_i, f'_i)$ is evaluated each time, and the chain rule becomes a matrix product:

```python
def autodiff(liste, x):

    # function and its derivative
    f, fp = liste[0]

    # leave: evaluate f and f' on x
    # x is then a real number
    if len(liste) == 1:
        return f(x), fp(x)
    # otherwise rate f on children
    # x is then a tuple (Fi, Fpi)
    else:
        # autodiff on the children
        out = [autodiff(liste[i], x) for i in range(1,
         ↵ len(liste))]
        # list of children's values
        Fi = [t[0] for t in out]
        # list of children's derivatives
        Fpi = [t[1] for t in out]
        # case of a function of a single variable
        if len(out) == 1: Fpi = Fpi[0]
        # applies the chain rule
        return f(*Fi), np.dot(fp(*Fi), Fpi)
```

Then, we test on *G*:

```python
autodiff(myG,2)
```

which corresponds well to the numerical approximation of the derivative:

```python
(G(2+h)-G(2))/h
```

And, finally on *H*:

```python
autodiff(myH,2)[0] - H(2), autodiff(myH,2)[1] -
 ↵ (H(2+h)-H(2))/h
```

**i.** With JAX, all you need to do is import grad:

```python
from jax import grad
```

then the result is obtained trivially:

```python
grad(f)(0.4)
```

With PyTorch, it is a little more complex: you need to define **x** as a tensor with automatic differentiation enabled:

```
import torch
x = torch.tensor(0.4, requires_grad=True)
```

The next step is to evaluate the function f at x and then calculate the value of the derivative:

```
fx = f(x)
fx.backward()
x.grad
```

Note that numerical differentiation is very unstable on this example:

```
h=1e-14
(f(0.4+h)-f(0.4))/h
```

because the function contains a jump composed 50 times.


## SOLUTION 12.6   NEURAL NETWORK (!)

**a.** Simply use Matplotlib:

```
xx = np.linspace(0,1,100)
plt.plot(x, y, ".")
plt.plot(xx, f(xx))
```

We observe that the data follow the $f$ function, but with relatively high noise.

**b.** It is a matter of taking the sum of the squares:

```
np.sum((f(x) - y)**2)
```

**c.** The sigmoid is simply defined with NumPy:

```
def sigmoid(z):
    return 1/(1 + np.exp(-z))
```

and the derivative is calculated by hand:

```
def Dsigmoid(z):
    return np.exp(-z)/(1 + np.exp(-z))**2
```

Finally, the graphical representation of the sigmoid is a function that is 0 when $x \rightarrow -\infty$ and 1 when $x \rightarrow \infty$ and interpolates smoothly and increasingly between the two:

```
xx = np.linspace(-10,10,100)
plt.plot(xx,sigmoid(xx))
```

**d.** For **x** real, the function corresponding to $f_\omega(x)$ is given in Python by:

```python
def F(x, omega):
    a, b, w = omega
    return np.sum(sigmoid(a*x + b)*w)/len(w)
```

When **x** is a NumPy vector, one has to extend the parameter dimension and sum only on the first axis:

```python
def F(x, omega):
    a, b, w = omega[..., np.newaxis]
    return np.sum(sigmoid(a*x + b)*w, axis=0)/len(w)
```

**e.** The gradient of $f_\omega(x)$ with respect to $\omega = (\boldsymbol{a}, \boldsymbol{b}, \boldsymbol{w})$ is:

$$\frac{\partial f_\omega(x)}{\partial \omega} = \left( \frac{\partial f_\omega(x)}{\partial \boldsymbol{a}}, \frac{\partial f_\omega(x)}{\partial \boldsymbol{b}}, \frac{\partial f_\omega(x)}{\partial \boldsymbol{w}} \right),$$

with

$$\frac{\partial f_\omega(x)}{\partial a_i} = \frac{1}{n} w_i \sigma'(a_i x + b_i) x,$$

$$\frac{\partial f_\omega(x)}{\partial b_i} = \frac{1}{n} w_i \sigma'(a_i x + b_i),$$

$$\frac{\partial f_\omega(x)}{\partial w_i} = \frac{1}{n} \sigma'(a_i x + b_i).$$

In Python and for **x** a NumPy vector, the idea is to concatenate the gradients with respect to the three parameter vectors:

```python
def DF(w, omega):
    a, b, w = omega[..., np.newaxis]
    return np.stack([Dsigmoid(a*x + b)*w*x, Dsigmoid(a*x +
    ↪  b)*w, Dsigmoid(a*x + b)])/len(w)
```

**f.** The implementation of $J$ is immediate:

```python
def J(omega):
    return np.sum((F(x,omega) - y)**2)
```

The gradient of $J$ is given by:

$$J'(\omega) = \sum_{k=0}^{499} (f_\omega(x_k) - y_k) \frac{\partial f_\omega(x_k)}{\partial \omega},$$

which gives the following implementation, taking care to sum only on the last axis:

```
def DJ(omega):
    return np.sum(2*(F(x,omega) - y)*DF(x,omega), axis=-1)
```

**g.** First, a function implementing the gradient descent algorithm and monitoring the cost function every `monitor` iteration:

```
def optimize(DJ, omega, eta=0.01, steps=100_000,
↳ monitor=1000):
    print(f"J = {J(omega)}")
    for i in range(steps):
        domega = DJ(omega)
        omega -= eta*domega
        if i % monitor == 0: print(f"J = {J(omega)}")
    return omega
```

Next, we initialize the parameters randomly, then perform gradient descent to learn the parameters:

```
nb_para = 4
omega = rng.random((3,nb_para))
omega = optimize(DJ,omega, steps=20_000)
```

The cost function does indeed decrease, then stagnates at a level comparable to the cost $J_f$ of the function $f$. Note that the cost of the learned function is even slightly lower than that of $f$. Finally, to plot the data, the function $f$ and the learned function $f_\omega$:

```
xx = np.linspace(0,1,100)
plt.figure(figsize=(8,5))
plt.title("Learning a real function")
plt.plot(x, y, ".")
plt.plot(xx, f(xx), label='$f$')
plt.plot(xx, F(xx,omega), label=r'$f_{\omega}$')
plt.legend()
```

Even with a single layer of four neurons, the learned function is very close to the original function, as shown in Figure 12.8.

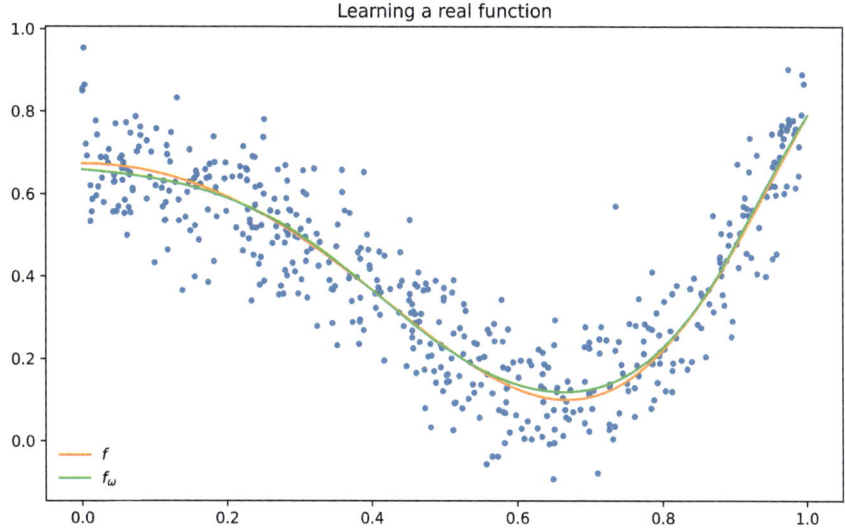

Figure 12.8   Representation of the function $f$, data generated by noisy evaluation of $f$, and learned function $f_\omega$ by a single-layer network of four neurons.

# CHAPTER 13

# Cryptography

Since the Caesar cipher, the cryptographic methods allowing to transmit secret messages evolved following the progress allowing to break them. The Vigenère cipher which is an improvement of the Caesar cipher will be studied and we will see how it is possible to break this encryption method. Then, the RSA encryption method which is one of the most used asymmetric cryptography methods today will be introduced.

Concepts covered

- Vigenère cipher

- greatest common divisor

- text import

- prime and pseudoprime numbers

- Fermat's little theorem

- Euclid's algorithm

- Miller-Rabin algorithm

- optimization by decorator

- asymmetric RSA encryption

DOI: 10.1201/9781003565451-13

# EXERCISES

## EXERCISE 13.1   VIGENÈRE CIPHER

The Vigenère cipher consists in choosing a key formed by a secret word (in capital letters) and to transform it into a vector whose elements are the positions of these letters in the alphabet. For example, "ASECRET" corresponds to (0, 18, 4, 2, 17, 4, 19). To encode a text (in capital letters, without spaces or punctuation) with the "ASECRET" key, you just have to shift the first letter by 0, the second by 18, the third by 4, and so on, and repeat in a loop. The details, especially the historical ones, are available on Wikipedia.

**a.** Write a function to_int(s) that transforms a character into its place in the alphabet and write the inverse function to_chr(i).
*Hint: See the documentation of the* ord *and* chr *functions.*

**b.** Write a function crypt(text, key) that encrypts text with the secret word key.

**c.** Write a function to decipher a text by knowing the key.

## EXERCISE 13.2   BREAKING THE VIGENÈRE CIPHER (!)

Charles Babbage was the first to break the Vigenère cipher. The idea is that three consecutive letters appearing several times in the cipher text are likely to be the result of encrypting the same letters of the message with the same letters of the key. This is even more likely with a group of four letters. Thus, the spacing between two same groups of cipher letters is a multiple of the key length. For example, if the repetition of one group is spaced 28 letters apart, then the repetition of another is spaced 91 letters apart, the greatest common divisor (GCD) of 28 and 91 is 7. So, it is likely that the key has 7 letters. Then knowing the size of the key, it is enough to base on the fact that the letter "E" is the most common in English. For this strategy to have a chance of success, the size of the text must be large enough.

**a.** Write a function to calculate the GCD between two numbers. Write another function to calculate the GCD between a list of numbers.

**b.** Visit the Project Gutenberg website (https://www.gutenberg.org/), choose your favorite English text and download it in "Plain Text" format. Write a function that converts the text to uppercase and strips it of all punctuation and other special characters.
*Hint: To convert a text to uppercase (converting accents) and remove all punctuation and other characters, it is possible to use the following function:*

```python
import unicodedata, re
def convert_upper(text):
    # convert to upper case
    text = text.upper()
    # convert accents
    text = unicodedata.normalize('NFKD', text)
    # delete special characters
    regex = re.compile('[^a-zA-Z]')
    text = regex.sub('', text)
    return text
```

**c.** Keep about of few thousand characters in the middle of the chosen text and encrypt it with a key. Then, write a function to determine the length of the key by looking at identical strings of three or more characters in the encrypted message. *Hint: First, form a dictionary with as key all occurrences of three letters and as value the positions of the occurrences. Then, determine the list of distances between the occurrences of three letters, then calculate the GCD of these distances. If this GCD is equal to 1 or is too small, then try again but with strings of more than three characters.*

**d.** Write a function to decrypt an encrypted message by returning the key. Try to decrypt the text of your favorite author with this function.
*Hint: To find the first letter of the key, it is necessary to calculate the number of occurrences of the 26 letters of the alphabet in the encrypted message that have been encrypted with the first character of the key. In principle, the letter with the maximum occurrence corresponds to the letter "E". It is then enough to do the same thing to find the other letters of the key.*

## EXERCISE 13.3   GENERATING PRIME NUMBERS

Most current encryption algorithms are based on the use of large prime numbers. The goal is to write a function to generate prime numbers. The first step is to generate a large random number, *i.e.*, having a certain number of bits. Then, a primality test allows to decide if this number is prime or not. If $\pi(n)$ denotes the number of primes less than or equal to $n$, then asymptotically $\pi(n) \approx \dfrac{n}{\ln n}$. For a number less than $n$ drawn at random, the probability that it is prime is about $1/\ln(n)$. For example, to generate a prime number of 1 024 bits (the minimum guaranteeing reasonable security at the moment), *i.e.*, of the order of $2^{1024}$, one must try $\ln(2^{1024}) \approx 710$ random numbers before finding one that is prime. Since all even numbers are clearly not prime, it is enough to test an average of 355.

**a.** Write a program to generate an odd random number of $k$ bits, *i.e.*, between $2^{k-1}$ and $2^k$.
*Hint: The fastest way to implement this is to use the bit operations explained at the address:* https://docs.python.org/3/reference/expressions.html#binary-bitwise-operations.

The simplest way to determine whether a number $n$ is prime is to try to divide it by all the integers $1 < d < n$. There are two reasons for not testing all $d$ between 2 and $n - 1$. The first is that it is unnecessary to try even $d$ greater than 2. The second is that there is no point in testing numbers larger than $\sqrt{n}$.

**b.** Write an algorithm isprime(n) to determine if a number is prime or not.

**c.** Write a function generate(k,primality) to generate a random prime of k bits with the primality test. Test with the primality test isprime. Is it reasonable to expect to generate a prime number of 1 024 bits with this algorithm?

## EXERCISE 13.4   GENERATING PSEUDOPRIME NUMBERS

The previous algorithm for generating primes being unusable for generating large primes, another approach, probabilistic, is advocated. A probabilistic primality test decides that a number is prime if it is prime with a very high probability. Such a number is called pseudoprime. Thus, a probabilistic test can be wrong and assume that a number is prime when in fact it is not.

The simplest primality test is based on Fermat's little theorem: if $n$ is prime, then $a^{n-1} = 1 \pmod{n}$ for all $1 \le a \le n - 1$. So, if we find an $a$ such that $a^{n-1} \ne 1 \pmod{n}$, then $n$ is not prime. Fermat's primality test tests $N$ values of $a$ chosen at random and if $a^{n-1} = 1 \pmod{n}$ for these $N$ values, then it declares that $n$ is probably prime. Carmichael numbers are not prime, but satisfy $a^{n-1} = 1 \pmod{n}$ for all $a$ prime with $n$. The prime Carmichael numbers are 561, 1 105, and 1 729. If $n$ is not a Carmichael number, then the probability that Fermat's test is wrong is $2^{-N}$. Choosing, for example, $N = 128$, we get a probability of being wrong of less than $3 \times 10^{-39}$.

**a.** Write a function implementing Fermat's primality test. Use this test to generate random pseudoprimes.

*Hint: See the documentation for the* pow *function for a quick implementation. If OpenSSL is installed on your computer, it is easy to check if a number is prime with the command* openssl prime 11, *for example, for 11.*

**b.** ! Improve the speed of the previous algorithm by first testing whether $n$ is divisible by primes less than 1 000 before applying Fermat's test.

Fermat's primality test allows to generate large pseudoprime numbers with a good probability of being right. The main problem comes from the existence of Carmichael numbers which are excluded from this probability. The Miller-Rabin primality test avoids this problem.

**c.** !! Understand and implement the Miller-Rabin primality test explained in detail on Wikipedia.

## EXERCISE 13.5   RSA ENCRYPTION

The RSA algorithm, from the initials of Ronald Rivest, Adi Shamir, and Leonard Adleman who invented it in 1983, is one of the most widely used asymmetric cryptographic algorithms still in use today. Asymmetric encryption allows an encrypted message to be transmitted to Alice without having to first transmit a secret key to Bob who encrypts the message. The creation by Alice of a public key is enough for Bob to encrypt the message and for Alice to decrypt it with his private key. There are three main steps in the RSA algorithm:

**Creation of the keys:** Alice wanting to receive a secret message chooses two very large prime numbers $p$ and $q$ which she keeps secret. She then computes $n = pq$ and the Euler's totient function $\varphi(n) = (p-1)(q-1)$ which counts the number of integers between 1 and $n$ which are prime with $n$. Then, she chooses an encryption exponent $e$ that is prime with $\varphi(n)$. The public key of Alice is given by the pair $(n, e)$. Finally, Alice computes the decryption exponent $d$ which is the inverse of $e$ modulo $\varphi(n)$, i.e., such that $ed = 1 \pmod{\varphi(n)}$. The private key of Alice is $(p, q, d)$.

**Encryption of the message:** To encrypt her message, Bob first transforms it into an integer $M < n$. The encrypted message is then given by:

$$C = M^e \pmod{n}.$$

**Decryption of the message:** The encrypted message $C$ is then transmitted to Alice. To decrypt it, Alice calculates:

$$M = C^d \pmod{n},$$

which is again the original message.

**Remark:** The prime numbers $p$ and $q$ must be truly random, otherwise it is possible to guess their values. The random numbers generated by the random module are generated with the Mersenne Twister algorithm. This algorithm is not considered cryptographically secure in the sense that an observation of about a thousand random numbers generated by this algorithm is sufficient to predict all future iterations. To generate cryptographically secure random numbers one would have to use the secrets module.

**a.** Show mathematically that the decoded message corresponds to the original message.
*Hint: If $a = b \pmod{\varphi(n)}$ and M is prime with n, then $M^a = M^b \pmod{n}$.*

**b.** Given $e$ and $\varphi(n)$, write a function bezout(e, phi) to determine $d$ such that $ed = 1 \pmod{\varphi(n)}$.
*Hint: Use the generalized Euclid algorithm to determine the GCD g between two numbers a and b and x and y satisfying $ax + by = g$.*

**c.** Write an algorithm generate_keys(length) that generates prime numbers $p$ and $q$ such that $n$ has length bits, then determines $\varphi(n)$, $e$, and $d$, and finally returns the public key $(n, e)$ and the private key $(p, q, d)$.

**d.** By choosing to encode each character on 8 bits, a string of length $\ell$ is written as a list $(a_0, a_1, \dots, a_\ell)$ with each $0 \leq a_i \leq 255$. This list can be converted into an integer $k$ in the following way:

$$k = \sum_{i=0}^{\ell} a_i 256^i .$$

Write a function `toint` and a function `tostr` allowing respectively to convert a string into this integer and vice versa.

**e.** Write a function to encrypt a text with a public key and another to decrypt it with the private key. To do this, we must make sure that the text is convertible to an integer less than $n$, otherwise we must split it into blocks and encrypt them separately.

## EXERCISE 13.6   BREAKING RSA ENCRYPTION (!!!)

Here is a public key:

```
(6811733848239939246347061235991894986724315842174369112785 7 |
  ↵   737867721430816193,
857480488977203937958496372891351154528533346142902255835 80 |
  ↵   93567068308193213)
```

and a message encrypted with this public key:

```
[5859647561950653479051376445728700975078277409165051349212 9 |
  ↵   616474442548981218,
631044417760880427913263809393347832875875146680949667 38 |
  ↵   945085681279666911085,
285540710278784283552200781062397242699759100959189201 53 |
  ↵   395621402277475101298,
597415013381133437536444387144537965290608679217567797 7 |
  ↵   732561102242895663472,
581127503299642996191371492487876120300295275466426173 07 |
  ↵   333293076280314987312,
357978894808801313614190441062853820242277522454839285 31 |
  ↵   935179265371527989598,
407206725649039531032575227039893283345874374156992335 14 |
  ↵   026965026895256621366]
```

**a.** Decrypt the previous message!
*Hint: It is probably necessary to choose a suitable algorithm, for example, using quadratic screens (QS, MPQS, SIQS).*

# SOLUTIONS

## SOLUTION 13.1   VIGENÈRE CIPHER

**a.** The `ord` function returns the Unicode number of the character and the `chr` function does the opposite. Thus, it is enough to shift so that the letter "A" corresponds to 0:

```python
def to_int(s):
    return ord(s.upper()) - ord('A')
def to_chr(i):
    return chr(i+ord('A'))
```

**b.** To encrypt, determine how much to offset each letter of the text, then make the offset:

```python
def crypt(text, key):
    out = ""
    for i,c in enumerate(text):
        # shift given by the key
        shift = to_int(key[i % len(key)])
        # add encrypted letter
        out += to_chr((to_int(c)+shift) % 26)
    return out
```

**c.** Decrypting is the same as encrypting but moving backward in the alphabet instead of forward, so you just have to modify the previous function:

```python
def crypt(text, key, reverse=False):
    out = ""
    for i,c in enumerate(text):
        # shift given by the key
        shift = to_int(key[i % len(key)])
        # reverse the shift to decrypt
        if reverse: shift = -shift
        # add encrypted letter
        out += to_chr((to_int(c)+shift) % 26)
    return out
```

To test that it works well:

```python
crypt("UNSUPERBEMESSAGECODE", "ASECRET")
crypt("UFWWGIKBWQGJWTGWGQUI", "ASECRET", reverse=True)
```

## SOLUTION 13.2   BREAKING THE VIGENÈRE CIPHER (!)

**a.** To calculate the GCD between two numbers:

```python
def gcd(a, b):
    if a == 0:
        return b
    else:
        return gcd(b % a, a)
```

To calculate the GCD of a list of numbers:

```python
def lgcd(lst):
    if len(lst) == 0:
        raise Exception("Impossible to determine a GCD")
    out = lst[0]
    for i in lst:
        out = gcd(out,i)
    return out
```

**b.** It is possible to download a file manually, then open it with the open function, but also to download it directly from Python:

```python
import urllib.request
# download Alice's Adventures in Wonderland by Lewis Carroll
#    in UTF8
url = "https://www.gutenberg.org/files/11/11-0.txt"
text = urllib.request.urlopen(url).read().decode('utf-8')
# convert to upper case without punctuation
text = convert_upper(text)
```

**c.** The following function constructs a dictionary with as key the set of three-letter words and as values the positions of the occurrences of these three letters, then computes the distance between the sets and returns the GCD:

```python
def length_key(text, mot=3):
    # constructs a dictionary with as key the set of words of
    #    three letters and as values the positions of the
    #    occurrences of these three letters
    d = {}
    for i in range(0, len(text)-mot+1):
        t = text[i:i+mot]
        if t in d:
            d[t].append(i)
        else:
            d[t] = [i]
    # lists the distances between occurrences of the same
    #    letters
    distances = []
    for p in d.values():
```

```
            # if more than one element
            if len(p) > 1:
                for i in range(0, len(p)-1):
                    distances.append (p[i+1] - p[i])
        # the key should be the GCD between the distances
        return lgcd(distances)
```

To test, we keep about five thousand characters of the text, we encrypt it, then we try to determine the length of the key:

```
# keep 5000 characters and encrypt
cypher = crypt(text[10000:15000], "ASECRET")
# determine the length of the key
for mot in [3,4,5,6,7,8,9,10]:
    length = length_key(cypher, mot=mot)
    print(f"mot = {mot} => length = {length}")
```

We can see that the key is of length 7 as expected.

On the other hand, if we try to determine the key of a randomly generated message, it does not work:

```
import random
random.seed(1234567)
# generate a random message of 1000 characters
random_text = "".join([to_chr(random.randint(0,25)) for _ in
 ↪   range(1000)])
random_cypher = crypt(random_text, "ASECRET")
# try to guess the length of the key
for mot in [3,4,5,6,7,8,9,10]:
    length = length_key(random_cypher, mot=mot)
    print(f"mot = {mot} => length = {length}")
```

**d.** The following function returns the dictionary of occurrences of each letter of the alphabet in the letters of the encrypted message that have been encoded with the i-th letter of the key:

```
def occurrences(text, length, i):
    # dictionary to contain the occurrences of the 26 letters
    d = {to_chr(i):0 for i in range(26)}
    # extraction of letters modulo the length of the key
    ↪   starting at i
    subtext = text[i:len(text):length]
    # establishes the statistics of occurrence of each letter
    for c in subtext:
        d[c] += 1
    return d
```

To decrypt the message without knowing the key, the first step is to determine the length of the key with the length_key function. To do this, we give ourselves a minimum key length key_min and a maximum word length word_max:

```
def decrypt(text, key_min = 5, mot_max=10):
    # loop to find the length of the key
    for mot in range(3, mot_max+1):
        length = length_key(text, mot)
        # if key found is long enough then probably right
        if length > key_min:
            print(f"The key is apparently of length {length}
                ↵  with words to length {mot}.")
            break
        # impossible to determine a key of sufficient length
        if mot == mot_max:
            raise Exception("Impossible to decrypt the
                ↵  message")
    # loop to find each letter of the key
    key=""
    for i in range(length):
        # calculate the occurrences of each letter
        occ = occurrences(text, length, i)
        # return the letter with the highest occurrence
        c = max(occ, key=occ.get)
        # this letter should correspond to "E"
        key += to_chr((to_int(c)-to_int("E")) % 26)
    return key
```

Try to decipher the cipher text:

```
decrypt(cypher)
```

which returns the encryption key. As the algorithm uses statistics on the most used letter in English, the key might not always be found correctly.

## SOLUTION 13.3   GENERATING PRIME NUMBERS

**a.** The following algorithm generates a random number between $2^{k-1}$ and $2^k$ and, if it is even, adds 1 to make it odd:

```
def rand(k):
    p = random.randrange(2**(k-1),2**(k))
    if p % 2 == 0:
        p = p+1
    return p
```

A faster way to generate such numbers is to generate a number of $k$ bits, then set the most significant bit to 1 to ensure that the number is greater than $2^{k-1}$ and set the least significant bit to 1 as well to ensure that the number is odd:

```
def rand(k):
    p = random.getrandbits(k)
    # apply a mask
    p |= (1 << k - 1) | 1
    return p
```

**b.** The idea is to test if the number is divisible by two and then to test the division by all the odd numbers from 3 to the root of the number. Note the use of the & operator performing the conjunction "and" bit by bit which allows to determine if a number is divisible by two in a slightly faster way than with modulo.

```
import math
def isprime(n):
    # test if n is even
    if n & 1 == 0:
        return False
    # test even odd number from 3 to sqrt(n)
    for d in range(3, math.floor(math.sqrt(n))+1, 2):
        if n % d == 0:
            return False
    # if no divisor found, then n is prime
    return True
```

**c.** Generates a random number of k bits, then tests if it is prime and repeats if not:

```
def generate(k, primality):
    while True:
        n = rand(k)
        if primality(n):
            return n
```

By testing, it is possible to generate 32 bits numbers quickly with this algorithm, but with 64 bits numbers it becomes very long (several minutes). It is therefore illusory to want to generate prime numbers of 1 024 bits with this algorithm.

## SOLUTION 13.4 GENERATING PSEUDOPRIME NUMBERS

**a.** Tests N values of a to determine if n is prime or not.

```
def fermat(n, N=128):
    # test N values of a
    for _ in range(N):
        a = random.randint(2,n-2)
        # if not prime
        if pow(a, n-1, n) != 1:
            return False
    # no a allowed to determine that n is composite
    return True
```

With Fermat's primality test, it is now possible to generate 1 024-bit pseudoprime numbers in a few seconds:

```
generate(1024, fermat)
```

**b.** To construct the list of prime numbers less than 1 000:

```
low = []
for i in range(3,1000):
    if isprime(i):
        low.append(i)
```

Here is a function that optimizes a primality test test by first testing divisibility by primes less than 1 000:

```
def optimize(test, low=low):
    def testopt(n):
        # if already in the list
        if n in low:
            return True
        # if even
        if n & 1 == 0:
            return False
        # if divisible
        for i in low:
            if n % i == 0:
                return False
        # otherwise other primality test
        return test(n)
    return testopt
```

This saves about a factor of two on speed of execution:

```
generate(1024, optimize(fermat))
```

**c.** Since the optimize function is used to optimize a test, it is possible to define the optimized miller_rabin function directly with a decorator:

```
@optimize
def miller_rabin(n, N=128):
    # write n-1 as d*2^r
    d = n-1
    r = 0
    while r & 1 == 0:
        r += 1
        d //= 2
    # test N values of a
    for _ in range(N):
        a = random.randint(2, n-2)
```

```
    x = pow(a, d, n)
    # if a^d = 1 then the test is false and continue with
    ↳  another a
    if x == 1:
        continue
    # determine if x^(2s) = -1 for s=0,...,r-1
    s=0
    while x != (n-1):
        # Miller-Rabin test finished so n is not prime
        if s == r-1:
            return False
        else:
            s += 1
            x = pow(x, 2, n)
return True
```

To test the Miller-Rabin algorithm:

```
generate(1024, miller_rabin)
```

## SOLUTION 13.5    RSA ENCRYPTION

**a.** First of all:
$$C^d = M^{ed} \pmod{n}$$

then, we distinguish two cases:

- if $M$ is prime with $n$, then since $ed = 1 \pmod{\varphi(n)}$, then $M^{ed} = M$ $\pmod{n}$;

- if $M$ is not prime with $n$, then $M$ is a multiple of $p$ or $q$ or both and we must distinguish these three cases:

    - if $M$ is a multiple of $p$ but not of $q$, then $M$ is prime with $q$. Since $\varphi(n) = \varphi(p)\varphi(q)$, then $ed = 1 \pmod{\varphi(q)}$ and thus $M^{ed} = M$ $\pmod{q}$. Since $M = 0 \pmod{p}$, this shows that $M^{ed} = M \pmod{n}$.

    - if $M$ is a multiple of $q$ but not of $p$, then the proof is identical to the previous case with $p$ and $q$ exchanged.

    - if $M$ is a multiple of $p$ and $q$, then $M = 0 \pmod{n}$ and the result is trivial.

**b.** Generalized Euclid algorithm that for given $a$ and $b$ returns $g$, $x$, and $y$ such that $ax + by = g = \mathrm{GCD}(a, b)$:

```
def egcd(a, b):
    if a == 0:
        return (b, 0, 1)
```

```
    else:
        g, x, y = egcd(b % a, a)
        return (g, y - (b//a)*x, x)
```

Then, the modular inverse is easily calculated:

```
def bezout(e, phi):
    g, d, _ = egcd(e, phi)
    assert g==1, "e has to be prime with phi"
    return d % phi
```

**c.** Generate the numbers p and q, then determine a possible e and compute the d with the bezout function:

```
def generate_keys(length):
    # generates two pseudoprime numbers p and q of size
    ↵   length/2
    p = generate(length//2, miller_rabin)
    q = generate(length//2, miller_rabin)
    # as such n should have length bits
    n = p*q
    phi = (p-1)*(q-1)
    # find e prime to phi
    while True:
        e = random.randrange(1, phi)
        if gcd(e, phi) == 1:
            break
    # determine d
    d = bezout(e, phi)
    public_key = (n,e)
    private_key = (p,q,d)
    return (public_key, private_key)
```

**d.** Converting a string to an integer is like converting a base 256 number to base 10:

```
def toint(string, bit=8):
    # base
    base = 2**bit
    # list of characters converted in the base according to
    ↵   their position
    lst = [ord(c)*base**i for i,c in enumerate(string)]
    # the sum is the representation in the base
    return sum(lst)
```

To go back, you have to decompose the integer in base 256:

```
def tostr(integer, bit=8):
    # base
    base = 2**bit
```

```
    # number of letters
    size = integer.bit_length()//bit+1
    # gives the list of letters by converting from the base
    lst = [chr((integer//base**i) % base) for i in
     ↪  range(size)]
    return ''.join(lst)
```

**e.** Function to encrypt with the public key:

```
def encrypt(public_key, text):
    # unpack public key
    n,e = public_key
    # length of the key
    bits = n.bit_length()
    # blocksize by encoding each character on 8 bits
    blocksize = bits//8
    # list of groups of blocksize letters
    groups = [text[i:i+blocksize] for i in range(0,
     ↪  len(text), blocksize)]
    # list of integers associated with groups of letters
    integers = map(toint, groups)
    # encrypt each integer in the list of integers
    return [pow(i, e, n) for i in integers]
```

Function to decrypt with the private key:

```
def decrypt(private_key, text):
    # unpack private key
    p,q,d = private_key
    # calculate n
    n = p*q
    # length of the key
    bits = n.bit_length()
    # blocksize by encoding each character on 8 bits
    blocksize = bits//8
    # decrypt each integer
    integers = [pow(i, d, n) for i in text]
    # convert integers to groups of letters
    groups = map(tostr, integers)
    return ''.join(groups)
```

To test that everything works well:

```
public_key, private_key = generate_keys(256)
cypher = encrypt(public_key, "This is a top-secret message!")
decrypt(private_key, cypher)
```

# Index